Risk Society and the Culture of Precaution

Also by Ingo K. Richter, Sabine Berking and Ralf Müller-Schmid

BUILDING A TRANSNATIONAL CIVIL SOCIETY *(editors)*

Also by Ingo K. Richter

DIE SIEBEN TODSÜNDEN DER BILDUNGSPOLITIK *(The Seven Deadly Sins of Educational Policy)*

Risk Society and the Culture of Precaution

Edited by

Ingo K. Richter
Emeritus Professor of Law,
Eberhard Karls University of Tübingen, Germany

Sabine Berking
Researcher, Irmgard Coninx Foundation, Germany

and

Ralf Müller-Schmid
Editor, Deutschlandradio Kultur, Berlin, Germany

Editorial matter, selection and introduction © Ingo K. Richter,
Sabine Berking and Ralf Müller-Schmid 2006. Individual chapters © the
respective authors 2006

All rights reserved. No reproduction, copy or transmission of this
publication may be made without written permission.

No paragraph of this publication may be reproduced, copied or transmitted
save with written permission or in accordance with the provisions of the
Copyright, Designs and Patents Act 1988, or under the terms of any licence
permitting limited copying issued by the Copyright Licensing Agency, 90
Tottenham Court Road, London W1T 4LP.

Any person who does any unauthorized act in relation to this publication
may be liable to criminal prosecution and civil claims for damages.

The authors have asserted their rights to be identified
as the authors of this work in accordance with the Copyright,
Designs and Patents Act 1988.

First published in 2006 by
PALGRAVE MACMILLAN
Houndmills, Basingstoke, Hampshire RG21 6XS and
175 Fifth Avenue, New York, N.Y. 10010
Companies and representatives throughout the world.

PALGRAVE MACMILLAN is the global academic imprint of the Palgrave
Macmillan division of St. Martin's Press, LLC and of Palgrave Macmillan Ltd.
Macmillan® is a registered trademark in the United States, United Kingdom
and other countries. Palgrave is a registered trademark in the European
Union and other countries.

ISBN-13: 978–1–4039–9695–4
ISBN-10: 1–4039–9695–4

This book is printed on paper suitable for recycling and made from fully
managed and sustained forest sources.

A catalogue record for this book is available from the British Library.

Library of Congress Cataloging-in-Publication Data

    Risk society and the culture of precaution / edited by Ingo K. Richter,
Sabine Berking, Ralf Müller-Schmid.
      p. cm.
    Includes bibliographical references and index.
    Contents: The politics of risk – Security policies: assessing and
mitigating risks – Health: limits of the precautionary principle – Problems
of risk communication.
      ISBN 1–4039–9695–4
    1. Risk – Sociological aspects. 2. Uncertainty. 3. Democracy.
4. Economic development. I. Richter, Ingo. II. Berking, Sabine. III. Müller-
Schmid, Ralf, 1967–

HM 1101.R5935 2006
302'.12—dc22   2006045374

10 9 8 7 6 5 4 3 2 1
15 14 13 12 11 10 09 08 07 06

Printed and bound in Great Britain by
Antony Rowe Ltd, Chippenham and Eastbourne

Contents

List of Tables and Figures vii
Foreword viii
Notes on the Contributors ix

Introduction 1

Part I The Politics of Risk

1 The United Nations – Stability through Diversity 21
   Klaus Töpfer

2 Risk in Hindsight – Towards a Politics of Reflection 28
   Sheila Jasanoff

3 Culture, Structure and Risk 47
   Charles Perrow

4 The Principle of Precaution in EU Environmental
   Policy 59
   Bärbel Höhn

5 The European Social Model from a Risk Management
   Perspective 72
   Günther Schmid

Part II Security Policies: Assessing and Mitigating Risks

6 Security Policies in a Transnational Setting 93
   Ursula Schröder

7 Terrorist Attacks at Nuclear Facilities 111
   Chandrika Nath

8 The Case of Oil Exploitation in Nigeria 131
   E. Remi Aiyede

Part III Public Health: Limits of the Precautionary Principle

9 The Responsibilities of the Media in Dealing with HIV/AIDS in Southern Africa   149
   Brett Davidson

10 A Comparative Analysis of Risk Perception Related to Human Health Issues   167
   Frederic E. Bouder

11 Risk, Precaution and the Media: The 'Story' of Mobile Phone Health Risks in the UK   184
   Adam Burgess

Part IV Problems of Risk Communication

12 The Rhetoric of Risk and Responsibility: Understanding the German Public Debate on EU Enlargement   207
   Matthias Ecker-Ehrhardt

13 Media Communication, Citizens and Transnational Risks: The Case of Climate Change and Coastal Protection   229
   Harald Heinrichs and Hans Peter Peters

Index   253

List of Tables and Figures

Tables

7.1	Events linked to media coverage of Sellafield	117
7.2	If you had to find out more about radioactive waste, how reliable and honest would you expect each of the following organizations to be?	126
8.1	Oil in the revenue profile of Nigeria	133
13.1	Composition of media sample for content analysis	236
13.2	Topics of media stories	237
13.3	Alarming vs. reassuring tone of media coverage	238
13.4	Risks mentioned in the media coverage	239
13.5	Cause-and-effect relationships of climate change	239
13.6	Conviction of the population that climate change is real and that it can be mitigated	240
13.7	Risks of climate change – dangerous, likely, great damage, uncontrollable	241
13.8	Options for mitigation and adaption	243
13.9	Survey results: mitigation and adaption	244
13.10	Sources of statements about risk in the media coverage	246
13.11	Sources and statements about coping measures in the media coverage	247

Figures

7.1	Keyword search for 'Sellafield' and 'Terrorism' in British print media from January 1998 to February 2005	116
12.1	Paradigm underlying the 'Yes' of the German debate on enlargement	214
12.2	Alternative paths to the paradigm on the economic conditions of democracy	216
12.3	Paradigm underlying the 'But' of the German debate on enlargement	218
12.4	Argumentative integration of the risks of enlargement	219
12.5	Argumentative integration of the conditions for enlargement	219

Foreword

This publication is based on the work of the Irmgard Coninx Foundation, Berlin, established in 2001. In cooperation with the Social Science Research Centre, Berlin and Humboldt University, the Irmgard Coninx Foundation set up the Berlin Roundtables on Transnationality to discuss current issues in international politics. As a result of the first years of research two volumes on new transnational developments are being published by Palgrave Macmillan.

Volume 1 deals with the question of whether a transnational civil society is possible and how far it could be a major factor in solving global and transnational conflicts. Volume 2 is on risk policy and discusses questions of risk prevention and precaution as well as risk communication.

The editors would like to thank Guido Albisetti (Berne), Antje Landshoff (Hamburg) and Sabine Richter-Ellermann (Berlin), members of the Academic Advisory Board of the Irmgard Coninx Foundation. The members of the Advisory Board provided invaluable support and encouragement for the endeavour of the Berlin Roundtables on Transnationality, among others Jürgen Kocka, president of the Social Science Research Centre, Berlin, Elmar Tenorth, former vice-president of Humboldt University, Berlin, Wolfgang van den Daele, Wolfgang Edelstein, Shalini Randeria, Folke Schuppert and Georg Thurn.

The editors would also like to acknowledge Paul Knowlton, who translated the Introduction and chapter 1; and Robin Benson, who translated chapter 4.

Notes on the Contributors

E. Remi Aiyede coordinates the in-country National Governance Assessment Survey for Nigeria at the Development Policy Centre.

Frederic E. Bouder is senior Research Fellow at King's Centre for Risk Management, King's College, London.

Adam Burgess is a senior lecturer in Sociology, Department of Social and Policy Sciences, University of Bath.

Brett Davidson is programme manager at the Institute for Democracy (IDASA), South Africa.

Matthias Ecker-Ehrhardt is senior researcher at the Social Science Centre, Berlin.

Harald Heinrichs is associate professor at the University of Lüneburg.

Bärbel Höhn is a MP for the German Green Party and was Minister for the Environment and Conservation, Agriculture and Consumer Protection for North Rhine-Westphalia from 1996 to 2005.

Sheila Jasanoff is Pforzheimer Professor of Science and Technology Studies at the John F. Kennedy School of Government at Harvard University.

Chandrika Nath is a scientific consultant at the Parliamentary Office of Science and Technology, London.

Charles Perrow is Emeritus Professor of Sociology at Yale University.

Hans Peter Peters is communication researcher at the Research Centre Jülich, Programme Group Human Environment Technology and honorary professor at the Free University of Berlin.

Günther Schmid is Director at the Social Science Research Centre, Berlin (WZB) and Professor of Political Economy at the Free University of Berlin.

Ursula Schröder is a researcher at European University, Department of Political and Social Sciences, Florence.

Klaus Töpfer is Director-General of the United Nations Office at Nairobi (UNON) and Executive-Director of the United Nations Environment Programme (UNEP).

Introduction

Ingo K. Richter, Sabine Berking and Ralf Müller-Schmid

The concept of risk has had quite an impressive career in recent decades: Going beyond the narrower field of economics,[1] it has found its way into the everyday vocabulary of politics and the media. However, it has doubtless experienced its greatest boom in the social sciences. We can distinguish two main strands of this development. Against the background of risk situations that have taken on global dimensions, be they environmental disasters, epidemics or poverty-related, Ulrich Beck was the first to note the development in industrialized societies of a 'boundary-breaking grassroots-democratic development dynamic', and summed up its structural causes under the term 'Risk Society'. For Beck and the sociologists who followed him, risk became the paradigm of a modernity that had become reflexive.

On the other hand, a methodologically autonomous field of research known as 'risk assessment' has emerged in empirical social studies.[2] Instead of seeking to analyse society as a whole, risk assessment involves an approach based on statistical analyses and mathematical models. The approach, oriented to behaviourist models of action and theories of rational choice, makes it possible to come up with a precise formulation of the phenomenon, one that is at the same time free of political and historico-philosophical generalizations.[3] This objectivizing approach, geared as it is to the natural sciences, has, however, led ultimately to a methodological schematism that has a noteworthy tendency to level out the specific differences between natural disasters, environmental damage and terrorist attacks. While the concept of a risk society must, in certain places, be seen as harbouring the danger of historico-philosophical exaggeration, risk assessment, with its narrowly empiricist approach, is always in danger of reducing the concept to the issue of the probability that an undesirable event will occur.

Dealing with uncertainty

One of the aims of the present volume is to work counter to a reductionist concept of risk in each of the aspects outlined above. The volume seeks to develop elements of a theory of risk that, while empirically oriented, disavows any attempt to narrow the concept down by empiricist means, and in doing so it focuses on the interface between a number of different disciplines. The volume therefore extends from sociological case studies and analyses of political and cultural contexts to more broadly conceived studies in the fields of democratic theory and political economy. It goes without saying that the volume is driven less by a uniform methodological approach than by a broad, diverse interest in current social developments.

What risks can be seen as characteristic for the conditions given today, and how is the way we deal with these risks shaping the development of a globalizing world? These questions are also bound up with a number of normative considerations: What conditions must a society meet if it is to open up a multiplicity of opportunities for its members without losing sight of the need to reduce risks to a minimum?

An approach as broad as this will inevitably have to come to terms with a relatively heterogeneous terminology. What, viewed in methodological terms, may at first appear to be a drawback will, it is hoped, be recognized for what it is: the expression of a complex state of affairs, the background to which will direct the reader's attention to the historical dimension of the issue under consideration. The day-to-day need to deal with hazards and risks has roots extending far back to the beginnings of human civilization. Unpredictable forces of nature and the outcomes of social interaction, hardly less calculable than the former, served from the very beginnings of civilization to place the individual and community under pressure to take action. The express purpose of the rituals of tribal cultures, namely to fence in the unpredictability of the spirits and the gods, is indisputable evidence of an insight, as mythically disguised as it may be, into the precarious condition of their members' lives.[4] The fact that, viewed scientifically, these are practices with questionable prospects of success is perhaps less remarkable than the anthropological dimension of the procedure involved. As Clifford Geertz notes in his still relevant study on the magical practices of 'primitive peoples': 'As soon as customary expectations fail to materialize, the idea of "sorcery" will not be far behind.'[5] For the ethnologist, the magic ritual is the expression not simply of a pre-scientific, unreflected approach to the world; it is, quite on the contrary, an expression of common sense, a behaviour oriented to practical certainty. The doings of the medicine man are justified not by any supposed omniscience on his part, but by the virtually irrefutable everyday insight that things more often than not turn out differently from how we thought they would. This casts the sorcerer in the light of a pre-figuration of today's insurance agent.

Long before the principles of the modern ensuring state[6] and a culture of precaution[7] had reached their present, highly differentiated institutional state, people sought to counteract on their own initiative the risks that threatened them. The technical possibilities given were at first limited, and magic proved to be a problematic concept. For spirits and gods turned out to be at least every bit as unreliable as the forces of nature they were supposed to protect people against. So what could be more logical than to seek an appropriate remedy precisely where neither supernatural whims nor the proneness to error of practical action cast doubt on the prospects of reliably managing adverse circumstances: in the sphere of ideas and theory?

The need for certainty and stability in a constantly changing environment is deeply inscribed in the origins of Western rationalism. In classical Greek antiquity theoretical reason became the main instrument used in the search for certainty. This search replaced the magical practices, but the horizon of the logos retained a clear consciousness of the imponderables and dangers that may thwart the plans of rational individuals. One of the most important tasks of the early sciences, of metaphysics and mathematics, was to establish a sphere of rational certainty outside the contingent zone of ordinary life, and to give this sphere, an essence beyond time, precedence over the transient and superficial manifestations of the phenomenal world. But even the otherworldliness of Plato's realm of ideas, as trouble-free as it may in principle have been conceived to be, was unable to obscure the fact that the everyday life of people in antiquity was highly vulnerable to disease, war and natural disaster. In retrospect, it may perhaps be said that this *conditio humana* has not changed much to this very day. While it is true that our knowledge of the laws governing nature and human life has grown immeasurably and high technology has given us a hugely differentiated set of instruments to solve an almost unimaginable range of problems, the development of our instrumental and cognitive capabilities has at the same time been accompanied by massive growth in the contingencies resulting from this development. It was the atomic bomb that cast the first blinding light on a development in whose perspective technical progress appears to be bound up logico-structurally with the concept of overkill.

Finally, the impressions left by the Cold War and Chernobyl served to ensure that the term risk society would, in the 1980s, come to be the standard self-description of Western societies. Since then the term has lost hardly any of its salience, and the dialectic of instrumental rationality and systemic contingency has defined the contours of recent studies on climate change and global warming. While hurricanes are devastating towns and entire regions, paying no heed to political boundaries or social and industrial development, we are again faced, more urgently than ever, with the question of whether and to what extent such disasters are themselves the indirect result of human intervention. To formulate the matter provocatively, we might even say that under the conditions of global risk, natural disasters are threatening to turn into cultural disasters.

Niklas Luhmann speaks of this connection in his system theory,[8] where he elaborates the difference between risk and danger: 'The potential loss is either regarded as a consequence of the decision, that is to say, it is attributed to the decision. We then speak of risk – to be more exact of the risk of decision. Or the possible loss is considered to have been caused externally, that is to say, it is attributed to the environment. In this case we speak of danger.'[9]

Following Luhmann, the present volume is concerned not with dangers or hazards, but exclusively with risks. Responsible handling of high technology, vaccination against diseases and epidemics, or transformation of a political system are seen as risky, since their positively connoted aims (chances) are inseparably bound up with negatively connoted consequences (damage). Speaking generally, we can be said to be concerned here with risks because, without exception, human action is constitutive for the structure of the processes under consideration. In contrast to the rational optimism typical of the Enlightenment, we no longer have any reason to expect *a priori* that rational planning and rational action will ever be brought into a state of contingency-free congruity.

Risk as opportunity

Viewed against this background, the question is why it is that people expose themselves more or less voluntarily to risks, even though they are aware that a risk may occur. It certainly is not the 'fascination of hazard', the 'thrill of risk'. Risks are simply not conceivable without the possibility of drawbacks – but not without the possibility of benefits either; indeed it can even be said that risks harbour dangers and chances simultaneously, and it is for these chances that we create and maintain risks, that we expose ourselves to risks. We build nuclear power plants to secure our energy supply, and in doing so we accept the risk of nuclear disaster. Again and again, wars have been fought for the spoils to be won, and the victims are inevitably part of the calculation. Warnings issued by conservative banking institutions against engaging in speculative operations do not stop unwitting investors from putting their money into risky deals, at least if wheelers and dealers have promised them good returns. Nor have the Pope's calls for sexual abstinence and restriction of sexual intercourse to the institution of marriage served to dissuade people from engaging in risky sexual behaviours.

It is astonishing that the risk-typical fact of the inextricable intertwinement of risk and chance, of profit and loss, drawback and benefit is so loosely anchored in the consciousness of most people. And even though there may be culture-specific differences at play here, it can be said that the concept of risk is associated primarily with the possibility of incurring loss, and less of making gain. This may well be bound up with risk communication. But still, it would be impossible to overemphasize the fact that social, economic, technological, scientific and political development, that social change and reform, that progress and innovation have always been associated with man-made risks, with risks that have opened up extraordinary opportunities, chances for which people have been willing to accept hardship, setbacks and possible losses. In the age of discoveries and of imperialism, the European powers spread their civilization across the whole world, took massive advantage of the chances involved, exacted great efforts from their own peoples and ruthlessly exploited the alien peoples they encountered. The victorious march of technology and the ideology of economic growth have opened up vast possibilities for all peoples, but not without burdening the world with the dangers of environmental destruction. Risk policy is always opportunity policy and hazard policy at once.

Risk distribution and injustice

The chances and dangers associated with the creation and maintenance of risks evidently affect people differently. Chances and risks are unequally distributed. Indeed it is often the case that while some people have all the chances, others are exposed to the dangers. In 1986 Beck, in his *Risk Society*, proceeded on the assumption that the risk society is no longer a class society in the narrower, Marxist sense, his reason being that today risk is distributed differently from wealth. Need, Beck notes, is hierarchical, smog is democratic in nature;[10] but he does not deny the unequal distribution of risk, especially at the global level. The benefits of high technology are expected to benefit the larger community, no doubt including in particular the companies that develop and use it; and while the dangers, the possible damage, likewise concern the larger community, they particularly affect certain groups, e.g. people living in the vicinity of a chemical factory, the consumers of a hazardous medicine, the passengers of a defective aircraft.

In recent years the unequally distributed consequences of HIV/AIDS have made themselves felt in especially dramatic ways. Drugs developed by Western pharmaceutical companies were affordable only to patients in rich countries, not for the higher-risk population groups in the world's poor countries. It took considerable political pressure, internationally concerted legal action and global media attention for developing countries to obtain discounts on HIV/AIDS drugs and establish the legal right to develop generic equivalents.

Risk and the rule of law

The legal system and the courts have traditionally had the task of finding compromises between the interests of 'risk groups', i.e. groups that create, maintain and live from risks, and groups that bear the burdens stemming from risks and are affected by their consequences. To cite an example, contemporary law recognizes the perpetrator principle, often referred to as the polluter pays principle, or PPP, under which those who create a risk must provide compensation in the event that the risk occurs. There is furthermore what is known as strict or absolute liability; in this case those who create a risk (and may profit from it) are held liable for its consequences, even if they cannot be shown to have been negligent. But these legal principles do not appear to compensate adequately for the unequal distribution of chances and risks when a risk in fact occurs. The traditional legalistic principles of compensation fail in particular in the face of the dimensions assumed by modern risks. The victims of the Thalidomide scandal – both the affected children and their parents – in the end received financial compensation with a court settlement. But this legal instrument failed when it came to compensating the victims of the Chernobyl nuclear disaster, and in view of the risks posed by genetic engineering all mechanisms of legal compensation would appear doomed to failure. Here politics is called upon to step in.

Governance and risk

Looked at historically, the question faced by politics has primarily been concerned with the tasks that the sovereign nation-state is expected to assume in view of the uncertainties and risks to which its citizens are exposed.[11] The state started out by assigning the police the job of providing security and stability; later, in the course of historical development, departments of trade and industrial supervision – originally a police function – were given responsibility, e.g. for approving certain technical plants and equipment and keeping businesses under constant scrutiny. In an analogous process, the state left it the task of achieving a balance between chances and risks to civil liability law.

In developed industrial societies, on the other hand, government activities were increasingly focused on industrial and infrastructure policy, a field which has come more and more explicitly to include risk regulation. In order to cover the risks to which large segments of its population, especially industrial workers, were exposed, the European-style welfare state adopted social insurance, and in particular accident and invalidity insurance, which entrepreneurs, i.e. those responsible for these risks, were obliged to fund.

In the last three decades of the twentieth century these traditional responses to the emergence and development of the 'risk society' came to be perceived as insufficient and unsatisfactory. The state seemed to have reached its structural limits. Police instruments were no longer up to the task of managing environmental risks, because far from being caused by individuals who could be held accountable, these risks were caused by the system of high technology itself. It is for this reason that sociologists like Charles Perrow in the mid-1980s defined the concept of risk against the background of the inherent proneness of systems to disorder or breakdown. In this context even the welfare state's hands seemed to be tied, because, far from being able to govern the redistribution of chances and risks, it was at best in a position to guarantee compensation for the damage resulting from risks. Beck developed his concept of sub-politics as a response to this challenge. In view of the failure of representative parliamentary politics to master it, the aim of the concept is to assign to civil society the task of dealing with risks.[12]

The transnationalization of risks

The dimension of the problem is made unmistakably clear by the existence of transnational risk situations that no longer pay heed to national boundaries. But sovereignty, and with it any chance of nationally legitimized intervention, ends at these very boundaries, and international communities do not necessarily intervene. Against the background of the risks posed by armed conflict, civil war and terrorism, there are considerable doubts as to whether a transnational civil society is really in a position to tackle these tasks, in particular if we imagine it operating fully independently of the state. This is why we often hear said today 'Bring the state back in', and why political scientists speaks of the 'new governance' and the 'ensuring state'. This call for a risk-adequate form of good governance is given particular emphasis in the contribution to the present volume by Klaus Töpfer, Director of the United Nations Environment Programme. His chapter is devoted to the role the United Nations can play in the face of environmental disasters, and it outlines a political strategy for the twenty-first century.

Since the 1980s, and in the specific micro-climate found in Germany, the Greens have played a pivotal role in setting the parameters of the debate on environmental policy. One of the party's protagonists, Bärbel Höhn, a former state environment minister, likewise deals with the issue of government action in the field of environmental risks, comparing the perspectives of environmental policy in Europe and the United States.

There is, however, little doubt that the criteria used to judge the state's effectiveness have shifted in connection with globalization. The state's initial priority was to guarantee the key internal and external conditions needed for society to be free. This it did with the help of the police and military. The welfare state was further expected to provide efficient infrastructure for economic and social activities, and in particular to promote industry and services, to create jobs and to guarantee a certain measure of social balance. But more, and different, things are expected of the 'risk society state': promotion of business and science in their efforts to participate in international competition, in development and cultural affairs, provision of opportunities to take part successfully in international competition, protection against the risks involved in such competition, and, generally, provision of some sort of blanket protection against hazards and risks.

This postmodern state is thus faced with a paradoxical task: it is expected on the one hand to guarantee individual freedom and prepare people for competition in a market society; and on the other, to protect people from the risks posed by the very same society. The state's effectiveness, the quality of its policies and the likelihood that a government will survive are judged in terms of whether or not a state manages to resolve this contradiction and come up with tangible successes in doing so.

Prevention and precaution

Viewed in ideal-typical terms, the 'risk society state' has two models on which it can base its approaches to dealing with characteristic risk situations: prevention and precaution.[13]

Under the first model responsible political action aims primarily to eliminate the causes of risks. A preventive peace policy seeks to resolve conflicts peacefully, and in the ideal case it may succeed in preventing any escalation of violence. This principle was dominant in particular in the age of superpower confrontation and the threat of nuclear war. In the same sense, preventive environmental policy will serve to avoid environmental risks, e.g. by imposing a ban on 'greenhouse gases' to ensure that we will have no hole in the ozone layer in the first place. Preventive health policy will eliminate, in the ideal case, health risks by banning toxic substances, e.g. by banning smoking in public places with a view to preventing lung cancer, which is caused by carcinogenic substances in tobacco smoke.

It is obvious that this principle is limited in practice. Many risks are due to economic and growth policies whose effectiveness is bought at the price of certain risks. A given measure of economic progress and the financial benefits it brings to the actors involved are, *per definitionem*, not to be had without such risks. Stated simply: It is not only the 'evil' tobacco industry that is to blame for lung cancer; plant and breeding technologies are also involved in the risk cycle, due in part at least to their unceasing efforts to boost food production and feed a burgeoning world population. In other words, government economic and technology policy has virtually no choice but to support risk-entailing innovations, the reason being that such innovations offer opportunities not only for the companies involved but also for the common good. The reason why preventive policy in the strict sense is narrowly limited in its options is that, viewed from the economic perspective, it is fraught with the risk of stagnation and recession.

But the aporias of statist preventive policy really make themselves felt when risk situations are created not by social activities that enjoy the support and protection of the state, but by state action itself. To cite an example, the Internet was initially developed for military purposes and was only later made available for civilian use, and has now become an indispensable instrument of global communication. In the face of criminal misuse of this technology by hackers, the state simply does not have the option of banning the communications network it itself has created. This notion seems as absurd and hopeless as the years-long attempts of the Chinese government to keep the Internet free of criticism of the regime there attest.

Now, we have a clear picture of how closely the concept of risk is bound up with the concept of a democratic, liberal polity. On the other hand, the effect of any extremely exaggerated preventive policy would be, at least implicitly, to embed citizens, in totalitarian fashion, in a dictatorial straitjacket of laws and regulations. Or should we ban smoking to prevent lung cancer? Should we ban unprotected sexual intercourse as a means of preventing HIV/AIDS? Evidently a liberal democracy must also grant its citizens freedoms that are bound up with risks for themselves, for others and for the wider community. Democracy can at best use prevention to limit risks, but never to rule them out completely.

Our second model for dealing politically with risks comes into play as soon as free and open societies are forced to take cognizance of the fact that their risk-prevention options are limited. If it turns out that we can prevent certain risks only by unreasonably restricting freedoms, the liberal view would be that in cases of risk we must aim to limit the scope of damage and to distribute the resulting burdens as equitably as possible. This is the point at which preventive policy slips into the garb of precautionary policy. Implementation of a policy of precaution is part and parcel of governmentality, even if the state delegates its precautionary duties, e.g. to those who cause or benefit from risks. Theoretically, there are a number of different precautionary measures that can be taken. A government may impose certain obligations on third parties, in particular on those who create risks that affect the whole of society. For instance, the operators of a nuclear power plant will as a rule be obliged to show that they have taken safety measures that appear suited to preventing a system failure that could release radioactive contamination into the environment – e.g. that the power plant in question will automatically shut down in the event of a hazardous incident. In the same way, operators of chemical factories can be required by law to dispose appropriately of their production wastes. The discussion over this so-called PPP principle continues to play a central role in European environmental policy.[14]

However, there are risk situations in which PPP is not applicable, e.g. the case of a contagious epidemic. But here too the state is called upon to take precautionary measures. In the ideal case there will be measures in place to protect the people affected, in particular evacuation and quarantine plans, sufficient transportation and hospital beds, vaccines, and so on.

We note here that while precautionary policy is geared especially to cases of technological and health risks, other risks are as good as inaccessible to the means outlined above. At present the long-term changes threatening our natural resources, e.g. climate change, can be countered, if at all, only by preventive measures, not by protective measures and emergency plans. There is no insurance against war, and neither educational nor compensatory measures provide any hedge against risk-laden policies. What is called for is a different, a less risky policy.

All in all, the distinction between prevention and precaution can be used to work out two main tendencies of political action. As a rule, a policy designed to prevent risk situations directly will be geared to a legislative practice that penalizes risky behaviour – e.g. smoking in public or failure to dispose safely of nuclear waste – with a view to rendering as unlikely as possible the event that a risk will occur (lung cancer, radioactive contamination). In a democracy a strictly legalistic preventive policy is conceivable only if it is at the same time able to ensure, by procedural means, that the norms and standards it sets are legitimate. Laws that ban risky behaviours are not perceived as a violation of personal and economics freedoms if it is beyond doubt that the norm or standard in question serves the higher-level end of the common good, and is not simply a product of state dirigisme. The examples of smoking and nuclear waste have served to illustrate the point that the actors affected and the interest organizations involved may differ substantially in their assessment of the legitimacy of such norms and standards, and this fact constitutes a challenge to any prevention-oriented policy, one that cannot be ignored.

On the other hand, precautionary policy seems to have the major advantage that it makes far less use of sanctions to influence the behaviour of individuals and is therefore more compatible with liberal freedoms. Preventive medicine in this sense implies that when people fall ill, precautions will have been taken to ensure that they receive the best possible medical care and that appropriate insurance cover is available to reduce as far as possible any social differentials in the quality of the care available. And in the case of a nuclear accident the state will be responsible for evacuating the population and providing sufficient radiation-proof shelters. However, this form of containing the consequences of a risk after it has occurred may on the whole been seen as justified only if the responsibility for it can be apportioned reciprocally between government and social actors. Health insurance can function only if it is at the same time used to anchor health as part of civic self-responsibility in day-to-day life. The same goes for the economy: State-imposed safety systems cannot work unless the operators of industrial plants are fully aware of their responsibility for the environment and contribute actively to preventing risks.

In a nutshell: preventive policy is an approach based on legislation and legitimized norms and standards, whereas precautionary policy is keyed to the self-responsibility of the actors involved as well as to provision of aid and support when a risk does occur.

Risk and communication

Even these few examples will have served to make it clear that while, in the end, prevention and precaution can be distinguished in terms of methodology, in practice both principles must mesh if an effective risk policy is to be established. One fundamental element of any such effective risk policy is communication.

One of the central theses of the present volume is that prevention and precaution – the essential aspects of state risk policy – can be implemented and evaluated only on the basis of communication. The idea makes sense intuitively, because in democracies government policy formation can only be understood as a communicative process.

The actors in a field of this kind, defined in terms of the rule of law and communication theory are, *per definitionem*, the people as the sovereign, organizations of representation like political parties, associations and the media, and the specialized fields of the public sphere, scientists, professionals, and experts.

As early as in the 1980s Jürgen Habermas pointed to the problems that result in connection with tendency towards a largely juridified risk policy. The persons in question are put under pressure not only by external emergencies but also by the need to specify formally what has in fact transpired. From today's viewpoint his diagnosis of the functional contradictions of the welfare state reads like a commentary on the aporias of humanitarian intervention, be it conducted by state organizations or NGOs: 'The implementing bureaucracies have to proceed very selectively and *choose* from among the legally defined conditions of compensation those social exigencies that can at all be dealt with by means of bureaucratic power exercised according to law. Moreover, this suits the needs of a centralized and computerized handling of social exigencies by large, distant organizations.'[15]

The question of when a risk occurs and who is affected by it cannot be answered as directly as statistically oriented risk models might lead us to assume. The diagnosis of a loss-entailing event is an integral element of a multiplicity of communication processes, not merely the set of empirical facts on which it is based. The contributions to the present volume also clearly illustrate how closely the perception of risks is bound up with communication processes. Risks cannot simply be described as facts of the natural world; indeed, they are the result of a complex process of interpretation. Within the social world, the actors are also interpreters of both their own actions and the circumstances that enlarge or restrict their scope of action. Viewed under the conditions of risk, this state of affairs entails a number of in part formidable problems for the democratic communication process which will be outlined only briefly in the present volume.

The trend towards scientification in particular is fraught with implications for the entire field of risk communication. Scientific findings are cited to substantiate both the objectivity and the impartiality of risk-related judgements. It is not only state bureaucracies that subscribe to this move towards scientific objectification of risk-conditioned emergencies; civil society actors and pressure groups that see themselves as the advocates of affected persons take the same approach.

The inherent dilemma becomes clear when a government, in part counter to the expressed interests of business and consumers, seeks to implement a ban on greenhouse gases designed to protect the ozone layer and to prevent climate change. An information campaign pointing to looming dangers for the environment would be unthinkable today without explicit recourse to substantiated scientific evidence. Data are collected and the possible consequences and damage are predicted. This process is accompanied by a critical public, which, as a general rule, will also resort to scientific evidence. Nor would the most spectacular Greenpeace actions, to name just one of the most publicity-minded NGOs, be imaginable without a number of scientific background assumptions, e.g. about the long-term development of sea levels under the influence of global warming.

What this means in effect is that, under the conditions of an open society, risk communication develops a structure that may at times be paradoxical. In the public arena, one expert opinion will often be presented to refute another; the objectification strategies pursued by the disputing parties will mutually obstruct one another. In an impasse of this kind the media have a key role to play. In cases in which the unfettered constraint of the better argument fails to produce a clear winner, professional opinion-makers enter the arena, and are sometimes in a position to give the public deliberation process a decisive turn.[16] Looking at the dimensions of the terrorist threat and the possibilities of containing it, the importance of climate change as well as for political counterstrategies, the danger posed by infectious diseases, and the need to address their social and biological causes – there is not one of these major issues of the present and the future that does not unfold its specific dynamic in the tension between scientific knowledge and media publicity. Sociology's task is not an easy one in this context. It can identify structures of risk communication and design critical standards to guide the actors involved. Niklas Luhmann always emphasized the fact that this form of precise observation is also a form of participation in social communication. The intention of the essays in the present volume is also to contribute to this end.

The authors and their chapters

The politics of risk

Klaus Töpfer emphasizes the importance of the United Nations for dealing with environmental risks whose destructive potential threatens to spill over beyond national boundaries. Starting from a disturbing diagnosis of the condition of the present state of global natural resources, Töpfer pleads emphatically for a strengthening of the United Nations, which not only has the longest history and the greatest know-how of all the international organizations but is also is the only one endowed with a mandate under international law.

Sheila Jasanoff's contribution is about risk and our attempts to cope with it in rational terms. Instead of conceiving risk as a property of the indeterminate future, she argues that risk is better seen as a projection of what we already know, have experienced and think we should control. In this respect, risk becomes a product of human imagination disciplined and conditioned by an awareness of the past. That awareness appears to be necessarily partial and selective: the past is not the same for everyone who experiences it. As historians have shown, the ruler's memory does not recall prior events in the same way as that of the subaltern; and litigation reveals that causes invariably look different from the standpoint of the victims and the originators of risky enterprises. Therefore, Jasanoff suggests a change of perspective: we may have to become better at developing strategies of collective recall and at seeing risk itself not only as the narrow concern of 'management' but as part of the wider preserve of 'governance'.

Charles Perrow likewise proceeds on the assumption that risk management must go hand in hand with renewed reflection on dealing with political power. A process of rethinking, though, will need to start at the national level if its aim is to come to terms with the challenge of international terrorism. Since risks are structurally conditioned, and not simply an occupational hazard of advanced industrial societies, and in view of the fact that the devastating effects of terrorist attacks can only unfold against the background of such structural weaknesses of open societies, we find ourselves in the midst of a race that democracy can win only if it reflects on and pays heed to its power-political strengths.

Bärbel Höhn argues from the perspective of political practice. She contrasts the concept of a culture of precaution with a policy whose one-sided orientation to cost-benefit considerations leads ultimately into a vicious circle. The neoliberal credo of profit-maximization appears to be inseparable from the risk of losing the vital natural resources on which we depend. By comparing, against this background, the different traditions informing American and European policy, she develops a pragmatic concept of sustainability that holds the balance between efficiency thinking and social utopia.

Günter Schmid raises the question of whether or not, under the conditions of a globalized economy, the European welfare state continues to have model character. The characteristics typical of the European model here prove to be extremely heterogeneous. The ideas of social security, low income differentials and full employment appear increasingly to have lost touch with European reality. For this reason Schmid argues in favour of a concept that goes beyond the social democratic 'third way' and breathes new life into the concept of social solidarity: collective risk management by state and society.

Security policies: assessing and mitigating risks

Ursula Schröder deals with the security situation since the end of the Cold War and the emergence of 'new wars'. In the light of transnational risks, a reactive policy of deterrence and defence has been replaced by a proactive policy of development, aid, cooperation, human rights and structural change that is supposed to eliminate terrorism, crime, poverty and environmental degradation. This policy is also committed to prevention and precaution; it is developing early warning systems and is not inclined to balk at intervention.

For a proactive policy of this kind, communication of its concerns is of fundamental importance – in particular communication with the populations of countries involved in military actions. Against this background, Schröder looks at examples of successful and unsuccessful conflict resolution and risk management, in Bosnia-Herzegovina, Macedonia, Afghanistan and Iraq.

Chandrika Nath deals with the risk of terrorist attacks on nuclear facilities. She analyses the basic conflict between the absolute need to inform the population and the need to keep state knowledge secret from terrorists.

In this special case, the actual risk assessment involved remains secret; but there is of course unofficial information and speculation. This lack of information serves to fuel speculation on the part of the media and NGOs, and this in turn casts a problematic light on the role they play in providing the public with the best possible information available. 9/11 has left its mark here; and since then the relationship between politics and the public appears to be marked to a high degree by secrecy and mistrust.

Remi E. Ayiede describes Nigeria's path to parliamentary democracy from the perspective of grassroots movements and NGOs. The crucial factor driving the broad-based civil society engagement at the start was above all the challenge arising from the risks that developed in connection with the exploitation of Nigeria's oil resources. The ruthlessness with which multinational oil corporations pressed forward called environmentalist organizations and other NGOs into action, and they in turn stepped up the political pressure on the Nigerian government. As the government slipped into the role of the errand boy of the oil corporations, democratic and environmentalist movements gradually coalesced together. Here responsible risk policy and democracy proved to be inseparable.

Public health: limits of the precautionary principle

Brett Davidson focuses on communication about HIV/AIDS in South Africa. For years President Thabo Mbeki has disputed the connection between HIV infection and AIDS. It was the media that ultimately prevailed against this political resistance, contributing to securing widespread acceptance of scientific evidence. The central task of the media in the case of HIV/AIDS in South Africa can be characterized as agenda-setting. By taking on information, awareness-building and education functions, the media also assumed a new responsibility, one to which they were not always equal. Davidson discusses the role played by the media in the conflict between freedom and responsibility of the press, journalistic ethos and reader interest, and truth and government propaganda.

Frederic E. Bouder looks at vaccination as a risk, or vaccination-related public communication and the effects it has on the behaviour of the population. In the UK an article in the *Lancet* on studies looking into a correlation between the triple vaccination against measles, mumps, and rubella on the one hand and autism on the other gave rise to a press campaign that led to a reduction of the vaccination rate from 90 per cent to 60 per cent. Not even a government counter-initiative was able to redress this situation, even though the correlation under study could not be proved scientifically beyond reasonable doubt. In a similar case, the French government was forced to discontinue a vaccination campaign against hepatitis B, even though the campaign, conducted with remarkable commitment, had already achieved a high vaccination rate of 75 per cent. The media played a key role here as well.

Adam Burgess is concerned with the health risk posed by the cellular telephone system, focusing in particular on the media campaigns conducted against the building of transmission antennae in the UK. Following on the heels of a test case in the United States, a movement developed there against the construction of transmission antennas in residential areas and near schools. An interplay developed between the social movements involved and the media. The media reported on actions directed against transmission antennas and the activists used the media, and especially the Internet, to advocate and propagate their cause. The British government responded to the public debate by appointing a fact-finding commission. Even though the feared health risks could not be proved beyond doubt on the basis of scientific evidence, the government was still forced to issue warnings, entirely in the sense of a culture of precaution, about the possible risks posed by cellular telephones, though without imposing any legal restrictions on their use.

Problems of risk communication

Matthias Ecker-Ehrhardt looks into the political risk of EU enlargement, focusing on the example of Germany. How could it have happened that Germany's 'Yes' to eastern enlargement turned into a 'Yes, but'? Ecker-Ehrhardt uses a methodological distinction to arrive at his central thesis: The linguistic foreground of the debate was at first dominated by a rhetoric that propagated the overcoming of communism, the unity of Europe and Germany's responsibility for the East as the primary goal. However, the cognitive background increasingly gained the upper hand. Concerns about security and prosperity in Germany and German influence in European politics led to an altered risk prognosis. The problem posed by a dwindling legitimacy of the EU's enlargement policy was only seemingly resolved by the agreement reached on reforms in the EU and an extension of the transition period for new member states.

Harald Heinrichs and *Hans-Peter Peters* provide an account of an empirical study they conducted on climate change and coastal protection on the German North Sea. They analysed the national and regional press and interviewed costal residents: How do they perceive the connection between climate change and flood risk, and how do they asses the measures taken by the governments concerned? One unexpected result of the study is that the population tends more to believe in natural than in social causes, despite the fact that, since Kyoto, the media have been constantly reporting on the greenhouse gas emissions and their effects. The majority of respondents are therefore also more in favour of precautionary measures, i.e. efforts to reinforce dams, than of preventive efforts based on reduction of greenhouse gases.

Notes

1. One of the earliest systematic studies stems from the time before the world economic crisis had led economic risks into entirely new dimensions: see Frank H. Knight, *Uncertainty and Profit* (Cambridge: Riverside Press, 1921).
2. See e.g., *Journal of Risk Research*, Official Journal of the Society of Risk Analysis (Abingdon: Carfax Publishing).
3. See e.g. Paul Slovic, *The Perception of Risk* (Trowbridge: Earthscan Publications, 2000).
4. See John Dewey, The Quest for Certainty, in *Dewey, The Later Works*, ed. Jo Ann Boylston (Carbondale and Edwardsville, IL: Southern Illinois University Press).
5. Clifford Geertz, Common Sense as a Cultural System, in *Local Knowledge: Further Essays in Interpretive Anthropology* (New York: Basic Books, 1983).
6. See François Ewald, *Histoire de l'État-providence* (Grasset, 1986).
7. The term 'culture of precaution' goes back to a study by Adam Burgess: *Cellular Phones, Public Fears and a Culture of Precaution* (Cambridge: Cambridge University Press, 2003).
8. Niklas Luhmann, *Risk: A Sociological Theory* (Aldine, 2005).
9. Ibid., p. 21f.
10. Ulrich Beck, *Risikogesellschaft* (Frankfurt am Main, 1986), p. 48.
11. See Sheila Jasanoff in the present volume, who makes a clear distinction between 'risk management' and 'governance of risk'.
12. See Beck, *Risikogesellschaft*, p. 300.
13. See Ursula Schröder in the present volume.
14. See Bärbel Höhn in the present volume.
15. Jürgen Habermas, *The Theory of Communicative Action* (Boston, MA: Beacon Press, 1987), p. 363.
16. See Nath, Burgess, and Bouder in the present volume.

Part I
The Politics of Risk

…

1
The United Nations – Stability through Diversity

Klaus Töpfer

The United Nations has six official languages: English, French, Spanish, Russian, Chinese and Arabic. As Under-Secretary-General, I can leave the choice of language up to you. Actually, though, this brings me straight to the issue at hand. UNEP has staff members from nearly 90 nations. They all speak English. I always say that this English is a kind of modern Esperanto, a new language. The outcome is a drastically reduced vocabulary that concentrates on words that have to cover a broad spectrum of meaning. However, the more we lose diversity, the greater the risk that people will understand different things by what we say.

We have conducted a study on the cross-links between cultural diversity and biological diversity. It is fascinating to see how closely they correlate. That is to say, losses of cultural diversity invariably entail losses of biological diversity, and vice versa. In this study we used language as an indicator of cultural diversity and found that there are at present approximately 6,800 languages in the world. Of these, 2,800 are on the red list of endangered languages. The red list in turn correlates well with the red list of endangered species. In other words, nature teaches us that wherever diversity dwindles, the consequence will be growing instability and risk for individual species.

Monostructures have never been stable systems of organization. I therefore see a great challenge in efforts to shape the world in an age of globalization – without paying the price of uniformity. People are unable to bear any longer being able to fall back on their cultural identity during the process of globalization.

We are currently experiencing something of the order of a renaissance of cultural identity. It is essential that globalization not be (mis)understood as Americanization; indeed, globalization must remain inextricably bound up with diversity, cultural identity and biodiversity. If this can be assured, we are very likely to achieve a high degree of stability.

Environmental policy as peace policy

This is the basis of the work being done by the actors and makers of environmental policy. It was when environmental policy came into being that people first started to think about the locational advantage implied by boundaries. A community's landfill is normally located on its borders, because in this case the neighbouring community is forced to bear half the inconvenience, smell, and so on. Economists refer to this phenomenon as 'beggar thy neighbour'. What I mean is that environmental policy has always sought to prevent the negative impacts of human activities, and the associated risks, from simply being shifted on to someone else. Thus the principal justification for the first worldwide environmental conference in Stockholm in 1972 was the attempt to identify the costs of consumption and production at an early stage and to seek to use this knowledge to come up with mechanisms for avoiding tension. One starting point was to integrate these costs from the very outset into our own cost calculations with a view to reducing the size of our ecological footprint and at the same time to harnessing this cooperation to the end of improving our overall use of natural resources.

I would like to illustrate this point with an example from the German state of Rhineland-Palatinate. I was active for some time as this state's environmental minister and was responsible, among other things, for the pollution levels of the Rhine. If one goes to the small towns and villages along the Rhine in Rhineland-Palatinate, one finds in each and every one of them a Tulla memorial or Tulla Street. Tulla is the engineer who was responsible for improving the Rhine's navigability, which he accomplished by straightening and considerably shortening the course of the river. The reason why memorials have been built in his honour is that his work served to reduce the size of Rhine's flood plains and to create more space for settlement. Straightening the Rhine was a classic 'beggar thy neighbour' policy, because increased flooding along the lower reaches of the Rhine was exacerbated by the solutions found to problems besetting its upper reaches. With the Rhine, as with many other rivers, we have now solved problems of this kind by working out and signing a dense network of treaties.

Today it is often claimed that attempts to confine water will prove to be one of the greatest dangers and risks to general stability.

The world population is growing by roughly 70–75 million a year. Growth of this kind clearly calls for an appreciable intensification of agriculture, in particular one that goes hand in hand with growing use of irrigation technology. A population increase of 1 per cent means a 2 per cent increase in water demand. Given the fact that water cannot be multiplied by natural means, problems will inevitably emerge. UNEP conducted a study to determine whether catchment basins have been a source of tension, conflict or even war, or whether in fact they have served more to motivate riparian states to engage in water-related negotiations. The latter has proved to be the case. We have produced an atlas of the world's water conventions for the catchment basins of rivers and aquifers. More than 200 such catchment basins are presently being used and managed by the countries bordering them. In other words, what we need are early warning systems that identify limitations on natural resources and thus ensure that 'early warning' translates into 'early action'.

Liability

What tasks are incumbent on the United Nations in this preventive peace policy, in avoiding risks to security? Our first task is to provide information about areas in which there is targeted investment and cooperation, and for which there is a need to produce conventions aimed at preventing risks stemming from the overexploitation of natural resources.

Let me refer here to the Rio Principles, which I find both fascinating and highly important. If we had to renegotiate them today, we would never reach the same outcome. Take, for example, Principle 15, which deals with the so-called 'precautionary approach' and the question of how best to deal with risks and knowledge that have not yet been scientifically established. When we buy a car, we know we could cause an accident and this is why we buy an insurance premium and insure ourselves against this risk. I believe that the question of liability and insurance-based solutions is going to come to play an increasingly important role in the world.

When I was German Federal Environment Minister, the government was always being criticized for the protracted periods of time needed for approval procedures. We were told: 'In Germany we don't even have draft documents on the table, while in America they are already producing them.' I then always offered my private sector partners a deal: 'You can expect me to adopt American approval procedures here as soon as you are willing to accept American liability practices.' And that was, as it were, the end of the debate on allegedly unduly protracted approval procedures in Germany, for these liability issues give rise to a very different kind self-interest in avoiding risks.

Implementation of environmental agreement

How can we ensure that conventions and agreements are complied with? Our major problem here is the fact that it is precisely in the environmental sector that we tend to produce paper tigers, and we are not really in a position to enforce environmental agreements. It is in this regard that we differ crucially from the World Trade Organization. The WTO does have an 'enforcement and compliance' capacity. If a country imports the wrong bananas, then it may find itself fined $900 million, and even have to pay that sum. We simply do not have this option.

So now we have a discussion on the need to restructure my organization fundamentally in order to strengthen its legal position and bolster its power of enforcement. One proposal would be to make an 'organization' out of today's 'programme'. The difference is relatively large. We are directly integrated within the General Assembly and the general mechanisms of the United Nations. In other words, we receive no independent contributions and have to rely on voluntary contributions. An 'organization' has far more independence and a jurisdiction of its own. However, I am and remain convinced that the right way to resolve these problems is not to create a fully independent 'World Environment Organization'. Indeed, my hunch is that we will remain within the United Nations. Of course, we could always also create a new organization within the UN system. In his opening speech before the General Assembly in 2003, President Jacques Chirac of France called for the development of a United Nations Environment Organization (UNEO) which would remain within the UN system; its aim being to identify the need, to call for and to carry out concrete action – not as a mega-agency, but rather on the basis of agreements negotiated among our member nations.

Ecological aggression

The United Nations Charter begins with the words 'We, the Peoples', and, at the 2000 Millennium Summit, the Secretary-General presented, in my opinion, an excellent declaration that bears these words as its title. With this in mind, we have endeavoured, in the so-called Millennium Development Goals, to formulate recognizable risks and to call for appropriate programmes of action. The greatest risks we face are inevitably to be found where marked differentials impinge more or less immediately on one another. It is much like plate tectonics. You can be certain that sooner or later there will be an eruption or an earthquake. Over the past years and decades, the rift between poor and rich has not only not narrowed, it has widened. And far from benefiting the poor, globalization and global trade have served to make the rich richer. This is not ideology, these are facts.

We asked ourselves at UNEP what our world would look like if it were mapped not in terms of geographical aspects and boundaries but rather with a view to gross national product. On a map of this kind Africa would virtually cease to exist and India would shrink drastically. If one asked, using the same method, how the world would look if it were viewed in terms of the number of children under 15 years of age, we would come up with a nearly reversed mirror image. Placed side by side, the diagrams used to illustrate this state of affairs make more than plain the risks I have addressed. Our world cannot be stable and risk-free. Today we are ever more aware that one reason for these differentials lies in the fact that the rich nations pass their environmental costs on to the poor countries. I refer to this as ecological aggression because we are in effect massively straining the vital resources of other countries, and this is bound to exacerbate existing tensions.

An article that appeared *Nature* in 2004 states, in the form of a forecast, that the main reason for the continuing dramatic decline in biodiversity is climate change. Climate change is assumed to entail the loss of one million species, the reason being that the capacity of ecosystems to adapt is unable to keep pace with the process of climate change. I had the opportunity to be one of ten environment ministers invited by the Norwegian environment minister to spend four days in the Arctic region. Travelling with an icebreaker, we advanced as far as the 82nd parallel. I was appalled at the massive changes that the process of climate change had already caused and shocked to see the threat faced by the flora and fauna characteristic of this ecosystem. Each year we lose 3.5 km^3 of ice in the Arctic. A study on the development of glaciers in Bhutan and Nepal arrived at a similar conclusion. These are not scenarios of the future, they are concrete phenomena that have been observed and recorded. Today we are forced to assume that there are already far more environmental refugees than refugees of other categories. Worldwide, more than 25 million people have had to leave their homes due to environmental disasters – because the bases of production have changed so much that these people are no longer able to remain at home.

We have always emphasized that nature, the environment, is the wealth of the poor. Rich people are more or less able to disconnect from nature. UNEP has presented an excellent study on the effects of climate change on the certainty of adequate snowfall in winter resort areas. We can prove that the snowline has moved upward by some 200–300 m. Many believe that snowguns can be used to solve the problem. After all, what do we need nature for? I was, for instance, somewhat dismayed to see that the first events of the Nordic ski championships were being held in the autumn of 2003 at the 'winter resort' of Düsseldorf on the Rhine. I also see that the Rhine-Ruhr town of Bottrop has built a so-called Skidome with 650 m of ski slopes.

Evidently, we can disconnect from nature. The poor do not have this option. Nature is their wealth, and if we take this from them, the result will be threats and tensions. These people have no perspective, and they are also beginning to understand that those who are – at least in part – responsible for their plight live elsewhere. In other words, to overcome poverty, the United Nations is forced to intervene with projects, legal instruments, conventions, investments and implementation controls.

Environment and development: the role of the United Nations

Environment and development must go hand in hand. Without a sound environment, poverty cannot be eliminated. In an age in which the need for multilateral solutions is being addressed as directly as it is today, it is of enormous importance to have a personality like Kofi Annan as Secretary-General, a leader who has the credibility required to push for multilateral decisions that – even in ongoing conflicts – lose sight neither of international law nor of human rights. Some 60 years ago we witnessed the birth of the United Nations from insights and perceptions stemming from the Second World War, and for this reason we need to see as our main priority this notion of peace, of the need to reduce tensions, risks and conflicts. I see our task not only in the provision of blue helmets. We have to work to ensure that as our knowledge grows, we point clearly to the risks involved in using this knowledge. In discussing modern technologies such as genetic engineering and nuclear technology, I usually revisit Dürrenmatt's play *The Physicists*, in which one finds the fine passage: 'We cannot roll back existing knowledge to the Nirvana of non-knowledge, we must assume social responsibility for it.' The prerequisite for this is that this information, this knowledge, be made accessible to the broad population, as stated in

Article 15 of the Aarhus Convention. In this regard I am convinced that the United Nations is more urgently needed than ever before to come to terms with risks in this world. Without this multilateral governance system, we are bound to return to a law of the jungle. I do not believe that this would be wise.

We must also point out that in coming to terms with risks we are well advised not always to seek grand, holistic solutions but also to look to partial changes. I am a follower of the great theorist Karl Popper, and I have always been of the opinion that healthy pragmatism cannot fail to lead to improvements. In a democratic system it must be possible for new majorities to alter the decisions adopted by old majorities. It has been noted, aptly, that the modern nation-state is too small for the big things and too big for the small things. What we see worldwide is not a grand world government but rather a process of oligopolization. The world's oligopolies are coming more and more to negotiate among themselves instead of under the auspices of the United Nations. In every oligopolistic system the great risk must be sought among those who are not oligopolizable and slip through the cracks. This is where the United Nations has its great responsibility, not as a global government but rather as an intermediary that attends to the needs of those who otherwise have no voice or cannot make themselves heard. The fact that we have more and more counter-groups, like those attending the WTO negotiations in Cancun, is evidence that we need to develop these governance systems and use them to reduce risks in multilateral decision-making.

2
Risk in Hindsight – Towards a Politics of Reflection

Sheila Jasanoff

Twice, barely eight months apart, first in Asia in December 2004 and again in the United States in August 2005, water has played havoc with hundreds of thousands of lives. The two disasters, a tidal wave (or tsunami, as the Japanese call it) and a hurricane, occurred worlds apart – spatially, socially, culturally, economically, in their apparent causes and impacts, and in the political and humanitarian responses that each evoked from governments and others. But there were profound similarities as well. Both were 'natural' disasters, in that each originated in the workings of nature beyond the bounds of human control, but both revealed the interactivity of the natural and the social in producing death, dislocation and damage on an unprecedented scale. Both also drew attention to the fact that we often learn most about risk – who is at risk, for what reasons and to what degree – only in hindsight, after risk transmutes into actual harm. Reflecting on what was like and unlike in these two events allows us to see more clearly how far we have come, as enlightened societies, in attempting to manage the risks of modernity. What techniques have we evolved for predicting the evils that lie ahead, and how effectively can our powers of prediction cushion us against the harms that seem continually to befall segments of innocent humankind?

Like the other contributions to this volume, this chapter is about risk and our attempts to cope with it rationally. But whereas risk is normally seen as a property of the indeterminate future, a dark foreshadowing of things to come, I will argue that risk is better seen a projection of what we already know, have experienced and think we should control. In this respect, risk is a product of human imagination, disciplined and conditioned by an awareness of the past. That awareness, moreover, is necessarily partial and selective: the past is not the same for everyone who experiences it. As historians have shown us, the ruler's memory does not recall prior events in the same way the subaltern does; and litigation reveals that causes invariably look different from the standpoint of the victims and the originators of risky enterprises. Even science, modernity's most reliable identifier of causes, narrows the possible channels of explanation in the process of illuminating them.[1] It follows that, to enhance a society's capacity for understanding risk, we may have to widen our analytic horizon to take in more than the predictive sciences. We may have to become better at developing strategies of collective recall and at seeing risk itself not only as the narrow concern of 'management' but as part of the wider preserve of 'governance'.

I am not alone in seeing risk as memory bumped forward. To some degree, the entire enterprise of technical risk assessment arose from an ability to deploy history instrumentally, through gradually increasing sophistication with what the philosopher Ian Hacking has called the 'taming of chance'.[2] The future, as inhabitants of the twentieth century gradually recognized, was not deterministic but governed by probabilities. Chance itself, however, was often not random, but amenable to more or less reliable forms of prediction, and knowledge of the past was crucial to the accuracy of many forecasts. The actuarial tabulation of recurrent bad events, from common illnesses and disability to accidents and financial ruin, and the estimation from these numbers of risk in individual lives, formed the basis of the insurance industry, one of modernity's signature social achievements. If in every life some harms are bound to occur, then insurance guarantees that no one has to face them alone and destitute, without resources to compensation for injury and loss.

The latter half of the twentieth century, however, gave rise to risks that escape humanity's powers of prediction, in part because there is little or no direct historical memory to fall back on in evaluating them. Some are so infrequent, distant in time or causally complex that they are literally, as well as figuratively, incalculable; others arise through historically contingent human behaviours that no one, it seems, could have imagined, let alone foretold. Yet such risks may be catastrophic, at the level of individual as well as communal life. It is hard to date the precise moment of emergence of incalculable risks, but a turning point may be the explosion of the first nuclear weapon at the Trinity test site in the New Mexico desert on 16 June 1945. That initial public demonstration of the results of the *Manhattan Project* alerted the world to the possibility of total annihilation. It is said that, moments before the blast, the eminent physicist Enrico Fermi 'began offering anyone listening a wager on "whether or not the bomb would ignite the atmosphere, and if so, whether it would merely destroy New Mexico or destroy the world" '.[3]

Since then, not only the continued threat of a nuclear holocaust but a succession of more or less devastating natural and human-made disasters have kept alive the spectre of essentially incalculable, and hence uninsurable, risks. How to govern *these* – not simply manage them – has emerged as one of the greatest technical and political challenges of the early twenty-first century.

Below, I reflect on the political dimensions of that problem. How should risks, especially incalculable ones, be governed? Four sets of difficulties have proved intransigent. First, there is the task of assessing the likely impacts of unprecedented events, with imaginations limited by deep-seated cultural commitments to particular ideas of rationality and particular ways of seeing and knowing risk. Second, there is the problem of identifying who is at risk from what, with associated questions about whose vulnerabilities are important and whose voices should count in setting priorities. Third, there are the unequal distributive consequences of risk, creating dilemmas about how to allocate scarce preventive and remedial resources. Fourth, there is the problem of drawing appropriate lessons from the past, given the tendency of governing institutions to simplify the richness and ambiguity of human experience. Below, I approach these problems by first reviewing how they played out, or more accurately were downplayed, in the development of risk analysis as a favoured discourse of public policy. I then show how these problems manifested themselves in the two natural disasters with which I began this chapter. I conclude by outlining a more modest, experiential and inclusive approach to dealing with the recurrent obstacles to democratic risk governance, using what I term 'technologies of humility'.[4]

The calculus of control

In the latter half of the terrifying twentieth century, risk became a major concern of governments. While (or possibly *because*) they had not been able to control the slide into two catastrophic world wars, advanced industrial states felt increasingly compelled to provide reassurances that they would be able to manage their citizens' destinies better in the future. In part, this meant that states had to show that the technologies their wars had done so much to promote – chemicals, biological agents and (later) genetic modification of living things, transportation, nuclear weapons, instruments of surveillance and mass communication – could be safely redeployed in the service of peace, prosperity and welfare. In part, it meant demonstrating that public officials have the capacity to foresee, and keep from materializing, new dangers on the horizon, dangers stemming from either the natural or the social world, or from their complicated interaction. In the past few decades, there have arisen on the agenda of risk prevention new environmental issues, such as deforestation, ozone depletion, species loss and climate change; new threats to peace from nuclear first strikes or, after 11 September 2001, global terror; new fears of pandemics, from AIDS to avian influenza; and new, seemingly ineradicable social 'diseases' such as poverty, scarcity and violations of human rights.

Demonstrations of the state's ability to manage risks, moreover, had to be made under changing sociopolitical circumstances that challenged the legitimacy of public reasoning. In democratic societies, citizens gained progressively more access to information and, through the expanding power of the media and the exponential growth of non-governmental organizations (NGOs), they also became better able to hold state authorities accountable for errors of omission and commission. With increasing wealth, more was at stake in getting things wrong, and the successes of modern technology bred growing expectations about what could be controlled and should not be left to chance. Demands for transparency grew, as publics asked to look behind the decisions of both elected and appointed rulers, and to question their proffered justifications. At the same time, the indeterminacy and complexity of many novel risks, and their refusal to stay within neatly drawn geopolitical lines, taxed the power of states to offer credible prediction and convincing remedies.

The vast and growing literature on risk chronicles the profound resulting transformations in the complexion of contemporary social and political life. To begin with, risk has shifted its locus almost imperceptibly from being a principally managerial problem to one that is also deeply political; put differently, risk management is now seen to be only one aspect of the broader enterprise of risk governance. Risk management was traditionally considered a domain for experts. Risk governance by definition requires the involvement of citizens and their political representatives. Whereas management entailed mainly the tasks of identifying and controlling risks, often associated with singular causes, governance takes as its purview the complex environments in which risks originate and concerns itself with explicitly political choices: for example, how to consider trade-offs among competing risks; how to communicate with affected (and disaffected) publics; and how to integrate demands for social justice into analytic processes once seen as largely technical.

Many changes converged to bring about this shift, changes occurring in domains of knowledge as well as of social organization and behaviour. On the side of knowledge, a key development was the emergence and spread of new formal techniques of risk analysis – aimed at disciplining the incalculable through advanced forms of calculation. Once the preserve of actuaries and insurers, risk assessment became at the end of the twentieth century an indispensable instrument in the armoury of governance. Mathematical models generating estimates of probable harm were developed for contexts as diverse as tracking the dispersal of chemicals in the environment and their effects on people or ecosystems, tracing the spread of introduced genes from one organism to another, measuring the likelihood of accidents at industrial facilities, and calculating the effects of global mean temperature rise on pathogens or storm surges. The proliferation of these methods raised new questions for governments: Whose knowledge should states rely on? How should policy-relevant knowledge be certified as reliable? How should expert judgements be made publicly accountable? How should conflicts among experts, interest groups and, indeed, national governments be resolved?[5]

On the side of social transformation, theorists focused, in the first instance, on the disruptive effects of risk on pre-existing structures and practices. Social analysts often took the fact of risk for granted; for them, it was the distribution and consequences of risk in society, and its disturbing implications for human autonomy and solidarity, that required investigation or explanation. In one sweeping treatment of the subject, the German sociologist Ulrich Beck famously proclaimed that risk had displaced class and other economic and social variables as the primary organizing force in modern life.[6] The 'risk society' Beck described seemed no longer able to rely unproblematically on science, the cornerstone of reason in public life since the Enlightenment. Through a process that he termed 'reflexive modernization', Beck argued, science not only provides persuasive support for the detection and assessment of risks, but, through its very proliferation supplies resources for scepticism and critique that prevent the construction of stable rationalities to support risk reduction. Others called attention to the relations of inequality that inevitably accompany the creation and dispersal of technological hazards, and their disproportionate impact on the economically marginal and socially disempowered.[7] Still others explored the troubling implications for democracy, as risk discourse and practice come to be dominated by powerful cadres of experts, allied with public and corporate policymakers, in the new, opaque formations of the regulatory state.[8]

By contrast, defenders of modernization and of its scientific and technological drivers played down the difficulties of both calculation and democracy, and constructed their own implicit models of what is wrong in contemporary risk societies. Alvin Weinberg, the influential director of America's Oak Ridge National Laboratory, sounded an early note of alarm with his much-cited observation that policymakers were turning to scientists for answers to essentially unanswerable questions, thereby defining a potentially ungovernable domain of 'trans-science'.[9] But, aided by growing computer power, the technocratic advocates of risk governance quickly regained confidence in the possibility of accurate prediction and control, even for seemingly incalculable risks, so that innovation could proceed, publics be reassured and limited resources be efficiently deployed if risks escaped managers' control. From the technocratic standpoint, the dominant problems are to find the right experts and produce the best knowledge. Most risks worth worrying about, the technocratic perspective holds, can be estimated on the basis of solid data, good models and (in an increasingly popular phrase) 'sound science'. Likewise, the credibility of risk assessments can be guaranteed through the careful separation of science from politics and through impartial peer review.[10] In this way, experts should be able to produce meaningful hierarchies of risk and make sure that the most significant ones receive the most attention.[11]

Deviations between expert and public perceptions of risk now surfaced as a source of perplexity and alarm. Why do publics often respond negatively to risks that experts deem negligible? In looking for answers, expertise in risk analysis became coupled with tacit theories of public opinion formation. If publics worry unduly about small risks, many concluded, it must be because of technical illiteracy and poor understanding of science, superstition, media hype or manipulation by political interests.[12] The study of public risk perception itself emerged as a new social science. Probing into the systematic variances between lay and expert assessments of relative risks, social psychologists identified several characteristics of risk that seem to enhance people's perceptions of danger: for example, novelty, involuntariness and lack of control.[13] Such research presumes that risk policy should not concern itself with distorting contextual variables that alter 'perception', but should look only at mathematically calculated probabilities of harm. It thus leaves intact the technocratic promise of being able to estimate risks with reasonable certainty and attributes the discrepancies between expert judgements and public perceptions to laypeople's inherent (and, by definition, distorting) cognitive biases. Rationality, for the expert assessors of risk, lies squarely on the side of scientific prediction and management. Concern about risks, especially those deemed small or negligible by experts, is correspondingly, and asymmetrically, labelled as biased or irrational, and treated as needing special cognitive explanation.

Some inroads into this binary framework of reason and unreason came from the field of organizational sociology. This literature displays the embeddedness of technological artifacts in matrices of organizational routines and practices that cannot be separated from the physical or biological pathways through which risks arise.[14] It is virtually a truism that human factors can contribute as much to risk creation as the inanimate features of functioning technologies. Accident investigations frequently ascribe the fault to a system's human implementation, through concepts such as operator or human error. Hence, ideas like Charles Perrow's 'tight coupling'[15] of complex technological systems or Diane Vaughan's 'normalization of deviance'[16] are now seen as integral to the design of safe technological infrastructures. Yet the very notion that the causes of failure can be separately attributed to human factors, especially the mistakes of individuals, exonerates the system's designers. It perpetuates the dream of perfect technological systems that would be failsafe if only their human components performed as flawlessly as their non-human parts. The human element of risky environments becomes, in this way of reckoning, simply another 'factor' to be incorporated into the framework of rational calculation and design.

None of this offers much comfort to proponents of democratic risk governance in a world undergoing rapid social, technological and environmental integration and change. Such governance would foster discovery and innovation in a world in which, all concede, zero risk is an unattainable ideal, but it would also take into account public preferences and concerns, as well as cross-cultural differences in wants and needs. The risk society discerned by Beck and others, however, offers few compass points by which to steer developments in science and technology in beneficial directions or to choose between alternative paths, each entailing some risks. In that sense, it is not a satisfying answer to the demands of what sociologists have also called *knowledge* societies: societies in which knowledge is the new form of capital, and its effective production and use are necessary for securing human welfare.

Still less promising are the analyses of risk experts who leave unexamined the very notions of 'risk' and 'expertise', who attribute failures of risk management mainly to human error, and who blame public fear and anxiety on laypeople's ignorance and incompetence in technical matters. Even organizational sociology's valiant efforts to restore the human to the domain of technical analysis has done little to focus attention on the political challenges of risk governance. As noted, a superficial reading of this literature can reinforce the belief in the perfectibility of machines that has proved such an impediment to accommodating wide-ranging social concerns about the production and management of risk.

Is it possible, then, to imagine a regime of democratic risk governance that allows for the possibility of change and yet makes room for the questions, doubts, fears and preferences of non-expert publics? Before attempting an answer, let us first look more closely at the two events that opened this chapter – the tsunami and the hurricane – focusing not only on the predictability of the events themselves, but also on the particular ways in which each one's consequences played out in the affected societies.

Unnatural disasters

On Boxing Day 2004, a massive earthquake measuring over 9.0 on the Richter scale occurred in the Indian Ocean, with its epicentre off the northern tip of the Indonesian island of Sumatra. The violent, minutes-long shaking produced a vast tidal wave or tsunami that radiated as far as Kenya and Somalia on the eastern coast of Africa. It left more than a quarter million dead, and many times that number bereft of shelter and livelihood, in Indonesia, Sri Lanka, India, Thailand and several other countries. The tragedy elicited an unprecedented global humanitarian response, though unequal it seems to the full dimensions of the tragedy.[17] While the scale of the relief effort generated its own tensions and controversies (was relief going to the right people, was it being used for the right purposes?), a second debate crystallized around the issue of preparedness. Why did so many die? Why were they not better forewarned?

Following the disaster, many commentators remarked that the Indian Ocean has no tsunami detection system of the kind that has existed for the Pacific Rim since 1949. Based in Hawaii, the Pacific Tsunami Warning Center uses seismic and oceanographic data to calculate the threat of a tidal wave following an earthquake and sends out warnings to regions likely to be affected. Analysts also noted, however, that such warnings could be effective only if they were communicated in a timely fashion to the populations at risk, and if the recipients were equipped to take preventive action on the basis of the warnings they received. In fact, tsunamis are relatively rare in the Indian Ocean, and human response capacities were correspondingly undeveloped. For example, in some places people did not recognize the warning sign of the sea retreating from the shore before it returned with killer force. Their curiosity aroused, both children and adults walked out to look at the fish left stranded on the shore, with fatal results. Communication broke down between and even within countries. Though several hours passed between the earthquake in Indonesia and the waves reaching India and Sri Lanka, people in those more distant countries were still not warned.

Although the waves took the lives of both rich and poor, loss and destruction were unevenly distributed. In the hardest-hit fishing communities, women and children died in larger numbers than men, who either were physically better able to survive or remained relatively safe far out at sea. According to an Oxfam report produced three months after the event, three to four times as many women died as men in surveyed regions in Indonesia, Sri Lanka, and India.[18] Aid to the survivors also divided to some extent along social and economic lines. Up to 2,000 European tourists, many holidaying in exclusive Thai resorts, lost their lives. A tiny fraction of the total death toll, they none the less received the greatest share of media attention in the disaster's immediate aftermath, contributing to a somewhat artificial sense of the disaster's global reach. Tourists' families benefited from the first efforts to identify the missing and the dead through DNA testing.[19] And allegations grew and persisted that relief efforts helped the rich more than the poor.[20] Harijans in India complained that they were among the slowest to receive aid.

The second disaster struck the United States, at the opposite pole of wealth and power from the countries hit by the Asian tsunami. On 29 August 2005, Hurricane Katrina made landfall near New Orleans, a city celebrated for its distinctive multicultural heritage, architecture and traditions. Built largely below sea level, New Orleans has relied on a system of levees to protect itself from flooding, but these were designed for at most a Category 4 hurricane and were overwhelmed as Katrina reached Category 5 strength in its rush from Florida to Louisiana. Instead of merely overflowing the barriers, flood waters breached the levees, causing breaks as much as 200 feet in length, with disastrous consequences for the city. Water from Lake Pontchartrain in particular poured through the breached 17th Street and Canal Street levee into the heart of the city. Before and after photographs showed block after flooded block of downtown New Orleans, with sheets of water where previously there had been a neat gridwork of street with houses in between.

Unlike the tsunami, which caught most of its victims completely unawares, Katrina came with prior warnings beginning as much as three days before. Indeed, the US National Weather Service predicted the possible breakdown of the levees, and Mayor Ray Nagin ordered a mandatory evacuation of the city on 28 August, a day before the hurricane struck. Not everyone, however, was in a position to heed the order, and once again the disaster's worst effects were felt most viscerally by the economically and socially marginal: the poor, the sick, the old and the infirm. Poignant stories included the deaths of hospitalized patients deprived of life-support systems, residents of old age homes, and people unable to escape from houses that were too low-lying to offer refuge from rapidly rising flood waters.

The disaster was one of the worst on record in American history. It left some 1,200 people dead, devastated a culturally vibrant city, and entailed by October 2005 an estimated $35 billion in property damage, surpassing by far the inflation-adjusted damage claims of $20 billion arising from Hurricane Andrew in 1992. New Orleans, in Katrina's wake, looked and smelled according to one account like a landfill, blanketed in 22 million tons of debris, beside the more than a million appliances, thousands of abandoned cars, and toxic materials from household chemicals to more hazardous substances like mercury.[21] More than a quarter of the city's housing stock, up to 50,000 houses by official estimates, was so badly damaged that it seemed impossible to salvage. In a city where nearly a quarter of the residents lived below the poverty line, it was unclear how many homeowners would have the resources to bear the massive cost of reconstruction.[22]

The political mess the disaster left was equally monumental. There were, to begin with, charges and countercharges between a Republican national government and Democratic state and local authorities about blame for failures of warning and, even more, of emergency response. An immediate target was Michael D. Brown, head of the Federal Emergency Management Agency (FEMA), whose lack of competence in disaster relief was widely cited as evidence of unacceptable cronyism in the Bush administration. Brown's public comments blaming the victims for some of their misfortune helped neither his nor his agency's image. And on a visit to the Houston Astrodome, where thousands of destitute New Orleans refugees had taken shelter, the president's mother, Barbara Bush, compounded the appearance of callousness by saying, 'And so many of the people in the arena here, you know, were underprivileged anyway, so this is working very well for them.'[23]

Geographically and culturally separated, the two disasters present some striking parallels. In each case, advanced technologies of prediction were actually or theoretically available, but were either inaccessible or incapable of getting people to take precautionary action. In each, how people behaved before and, above all, during and after the catastrophe influenced the extent and distribution of damage. In each, too, the disaster affected the more and less endowed – the rich and the poor, the old and the young, men and women – in dramatically different ways.

On the whole, people with more resources were better able to fend for themselves, both when warnings were delivered and after disaster struck. The losses in each case accordingly reflected not only the physically destructive impacts of earthquake, wind and water, but the unexpected entwining of these forces with features of human behaviour, technological capacity and social organization. In this sense the disasters, seen in hindsight, seem anything but 'natural'.

Following events of this tragic magnitude, it is common to ask whether, if the same thing happened again, we would be better prepared. That is the salient question policymakers have confronted in both the tsunami-devastated regions of Asia and the hurricane-torn city of New Orleans. Indeed, just two weeks after Hurricane Katrina, I was asked whether I would advise the construction of higher levees in New Orleans. After all, my interlocutor added, isn't this the sort of question that a scholar of risk from a leading university should be prepared to answer? What use is all that scholarship if it cannot inform decision-makers confronted with very practical problems how to choose between competing options? Should a tsunami detection system be installed in the Indian Ocean; should a drowned city be rebuilt on its old fault-lines?

There is no question that decision-makers face many such urgent questions in the aftermath of disastrous events; and it is important for them to make choices that are both responsive to sufferers' immediate needs and capable of minimizing future loss. In just over twenty years, profoundly difficult issues, technical as well as political, confronted decision-makers in India after the Bhopal gas leak, the former Soviet Union after the Chernobyl explosion, Britain after the BSE ('mad cow') epidemic, New York City after the 9/11 terrorist attacks, Indonesia after the tsunami, and New Orleans after Hurricane Katrina. One should not minimize the case-specific challenges of deciding how to clean up, make whole, restore safety and trust, and rebuild community and confidence after such cataclysmic risk-into-reality episodes.

And yet, the foregoing list, which could be indefinitely expanded, is itself instructive. Globally, there is no society that is not today a risk society. But risks of all kinds become apparent throughout the contemporary world only after they have materialized into harms. Given the range and severity of injurious events that can overtake us unawares, and the woeful shortcomings of emergency response despite decades of expert risk management, perhaps the important question is not how to predict events more accurately but how to ensure greater resilience if and when they occur. It is to ask how social capacities can be cultivated so as to ensure that bad events, when they happen, do less damage to human lives and solidarity.[24] The challenge, in other words, is to move away from a near-exclusive focus on causes and probabilities – on calculating the incalculable – towards a deeper understanding of the contexts within which injuries are experienced, and often exacerbated, with painful inequity.

Technologies of humility

The standard definition of risk used by policymakers worldwide is that it is the probability of harm times the magnitude of harm. This colourless definition performs two functions, both of which are essential from the standpoint of maintaining official credibility. First, it helps to naturalize risk – that is, to place it 'out there' in the real world, as a feature of that world's natural functioning. From this definition it is easy to draw the implication that 'zero risk' is unattainable; harms *will* occur, it is only natural, and the important thing for policymakers to know is the likely frequency of any particular harm, and how grievously it will affect those it strikes.

Second, the traditional definition provides a blueprint for what decision-makers need to know and measure in order to assess risks responsibly. In its widely circulated 1983 report on risk assessment, the US National Research Council set out a sequence of steps that together comprise risk analysis: hazard identification, exposure assessment, risk characterization, risk communication, risk management. Each step can be broken down into further sub-tasks that operationalize it; indeed, each has become a node around which new analytic techniques and new forms of expertise have crystallized over time.

Thus, highly technical methods of hazard identification, exposure assessment and risk characterization have developed around specific categories of risk. For example, today there are entire communities of experts dedicated to studying the environmental dispersal and human health effects of single pollutants like mercury and dioxin. Tsunamis and hurricanes along with many other natural hazards, such as volcanic eruptions, climate change, deforestation and biodiversity loss, have all given rise to bodies of expert knowledge produced by global networks of scientists. High-risk technologies, such as chemical plants and air traffic control, command their own specialist risk analysts. Since the threat of global terrorism emerged on the public agenda, it too has begun to generate new forms of expertise – for instance, the capacity to detect suspicious movements in crowds by means of computer models or trained human eyes. And few large regulatory bureaucracies could think of functioning without their in-house experts in risk-benefit analysis and, increasingly, risk communication.

The work of such experts is, of course, invaluable for decision-makers charged with protecting publics from risk. Accumulated expert knowledge represents an important kind of learning from the past – in this case, through science's powerful methods of gathering and building on its own prior successes. But as the litany of disasters recited above shows, there is no necessary correlation between the deeper scientific understanding of a particular risky phenomenon or system and the devastation experienced when the risk it presents mutates into reality. All our knowledge of tsunamis and hurricanes could not prevent stupendous damage in Sumatra and Louisiana, or even ensure prompt and effective assistance to those left stranded in each event's wake. In cases like Bhopal, BSE, Chernobyl or 9/11, expert knowledge of many relevant subsystems failed to produce the synthesizing forecasts that might have helped guard against those particular events. Despite all efforts at calculation, these realized risks arose in contexts in which the harms were, for all practical purposes, incalculable in advance. If they could not have been calculated, we may ask, could they nevertheless have been mitigated through improved governance?

That question requires us to turn on its head the analytic perspective so carefully developed through decades of effort to predict and control risks. For where management implies a top-down perspective, that of the manager in charge of a system, governance in democratic societies necessarily works from the bottom up – through the delegation of responsibility to rulers who are trusted to exercise power legitimately, on behalf of the people. The predictive stance relies on the manager's presumptively superior knowledge and expertise. Governance, by contrast, draws its strength from below, by aggregating communal knowledge and experience, preferences and concerns, which no science has brought under its control. Humility, not hubris, is the animating spirit of governance.

Recognizing the flaws and failures of traditional risk management does not mean, however, that one has to abandon analysis. Governance, too, can rest on its own approaches to systematic, analytic thought. Indeed, as I have suggested elsewhere,[25] it is possible to abstract from the case study literature on disasters, as well as from critical studies of risk analysis and policy-relevant science, four focal points for the 'technologies of humility' that can inform risk governance. These are *framing, vulnerability, distribution* and *deliberative learning*. Together, they address the questions that must be asked in seeking solutions to governance problems in any risky environment: what do we know about the risk and how do we know it; who is likely to be hurt; how will losses be distributed; and how can we reflect most effectively on our collective experiences of vulnerability and loss?

Framing

Policy scholars have recognized for some time that the quality of the solution to a perceived social problem depends on the adequacy of its original framing.[26] If a problem is framed too narrowly, too broadly or simply in the wrong terms, the solution too will suffer from the same defects. Was the primary problem in New Orleans that the levees were designed for hurricanes of lesser force than Katrina, or that building large parts of a city below sea level was a recipe for disaster? Did the focus on hurricane prediction distract people's attention from the massive relief efforts that would be required if the levees broke? Was the tsunami a failure of technology or of social systems (or both)? Will it help to install an advanced tsunami detection system in the Indian Ocean; or will the warnings generated by such a system fail to travel (as happened with the news of the tsunami) or to influence affected people's behavior (as happened with the warnings around Katrina)? Few policy cultures have adopted systematic methods for asking such questions about the dominant framings of risk, despite high-profile calls for doing so.[27] Frame analysis accordingly remains a critically important but neglected area of policymaking.

Vulnerability

Risk analysts have traditionally viewed at-risk individuals or populations as passive objects in the path of the risk to be characterized. People are seen as exposed to the risks that the manager wishes to control; their exposure is then assessed through techniques of formal quantification. One problem with this approach is that the risk manager's judgement is taken as the reference point for determining vulnerability, rather than the affected subjects' self-perceptions. Yet how people perceive their own vulnerability may be radically different in meaning and intensity from the perceptions of outsiders. As Barabara Bush's unthinking comment about 'underprivileged' New Orleans residents shows, outsiders may have believed that those communities had a lot *less* to lose than did members of the communities themselves. By contrast, it probably did not take a tidal wave to show the fishing families of the Indian Ocean region how precarious were their livelihoods and means of subsistence – or how completely their futures were tied to nature's fury or beneficence.

A second, related problem is that risk assessment tends to overlook the social foundations of vulnerability and to classify at-risk populations in accordance with supposedly objective physical and biological criteria.[28] In the effort to produce policy-relevant assessments, human populations are often classified into groups that are thought to be differently affected by the hazard in question (e.g. most susceptible, maximally exposed, genetically predisposed, children or women). These approaches not only disregard potentially salient differences within groups, but conceptualize individuals in statistical terms, as members of aggregates. This characterization leaves out of the emerging calculus of vulnerability factors such as history, place and social connectedness that may, in fact, play crucial roles in determining the resilience of human societies. As stories from the tsunami-affected regions showed, groups deemed similar on the basis of objectively measurable criteria coped very differently with disaster because of divergences in the strength of their communal knowledge and networks.

Through participation in vulnerability analysis, ordinary citizens may be able to correct some of these defects. They would in any event regain their status as active subjects in risk governance, rather than remain the objects of yet another distanced expert discourse.

Distribution

Perhaps the most striking similarity between the tsunami and the hurricane was that both disproportionately afflicted the poor. So notable were the socioeconomic disparities in New Orleans that David Ellwood, Dean of Harvard University's John F. Kennedy School of Government, said, 'It took a terrible hurricane, but the poor in America, who have languished largely unmentioned by politicians of both parties, are visible once again.'[29] Arguably, as the Oxfam analysis of discrepant gender-based impacts suggests, the tsunami played a comparable role in making visible the inequalities between women and men that were always there for any discerning observer to see. Such distributive effects, however, are not part of the agenda of classical risk analysis. To be sure, a law or policy may specify that the distribution of risk for different social groups needs to be taken into account, but in the absence of such mandates mapping inequality is not within the normal technical repertoire of exposure assessment.

In a regime of risk governance, by contrast, inequality would always be on the agenda because it is a factor that immediately and obviously affects people's responses to risk. Distributive issues would simply be part of the environment that decision-makers are responsible for understanding in their efforts to protect those most at risk. In deliberative settings, it would be flagged by those for whom risk policy is supposed to be made, much as claimants for environmental justice in an earlier era signalled their needs to regulators who had ignored the tendency of risks to cluster in poor and socially marginal communities. In a world where the task of dealing with risk focused on contexts as well as causes, and took into account subjective experience as well as objective measurement, it would not take a hurricane to make visible the plight of the poor.

Deliberative learning

The capacity to learn is constrained, as I have suggested, by the limiting features of the frame within which institutions reflect on their prior actions. Institutions see only what their governing discourses and practices allow them to see. Thus the framework of risk assessment continually reorients the expert learner's attention back towards prediction, with its emphasis on improved modelling and management of causes. It is hardly surprising, then, that so much of the immediate post-disaster discourse in our two cases focused on whether to build an Indian Ocean tsunami detection system or to raise higher the levees in New Orleans.

Experience, however, tells different stories to different people. It changes with the angle of observation. Even when the outward facts of a tragedy are more or less unambiguous (as they often are not) – what happened, who died, how many were hurt – its causes and consequences may nevertheless be open to many different readings. Just as there can be no single explanation for what caused the great wars of the twentieth century, so the failures of policy after major disasters cannot be attributed to unitary causes, lending themselves to inadequate patchwork remedies. The current design of most risk management institutions promotes the accumulation of expert knowledge and the revision of policy without challenging the manager's founding assumptions. In the shift towards risk governance, the aim should be to construct institutions of civic deliberation, through which societies can reflect on ambiguity and assess the strengths and weaknesses of alternative explanations. Deliberative learning, in this sense, may be messier in its processes and more modest in its expectations than expert practices of calculating risk; but it would be rightfully more ambitious in seeking to learn from the full extent of relevant experience and in building, on that basis, more resilient societies.

Conclusion

Humanity's attempts to grapple with risk in the twentieth century have turned out to be a task of protean dimensions. The concept at the centre of attention – risk itself – has proved to be as varied in its causes as nature itself, and as inevitable in its tragic consequences as the sufferings of the poor and the powerless. From early roots in accident and fire prevention, and natural hazard management, risk has grown to be a central preoccupation of public authorities seeking to justify their existence, as well as an organizing category of thought for modern societies. No one in today's networked world can fail to feel at risk from some, perhaps many, causes. Much of the work of regulation and even large segments of corporate enterprise are directed towards controlling risk in one form or another, whether in health care, hazardous industries, environmental protection, computer security, anti-terror activities or financial systems management.

Paradoxically, in what could perhaps be seen as a specific working out of reflexive modernization, the spread of risk in contemporary societies has undermined the very property that gave the notion its original power: namely, its amenability to calculation. Risk is the product of human imaginations bent on the taming of chance. Yet as modern humanity became more and more averse to leaving any aspect of its fate to be ruled entirely by chance, risk spilled out of the envelopes of measurement and prediction and became incalculable. In this process, as I have suggested, risk also escaped the control of expert managers and became a problem for politics: risk management, more particularly, gave way to the broader challenge of risk governance.

This fundamental shift requires a rethinking of the conceptual categories and analytic strategies with which human societies seek to cope with risk. Unlike management, governance cannot be conceived as a top-down process, the preserve largely of technocrats trained in specialized practices of prediction and control. Rather, like any political undertaking, democratic risk governance demands constant engagement with its authorizing public constituencies. It calls for a comprehensive mining of collective experience, geared towards building resilience rather than providing issue-specific prevention or relief. It also requires new approaches to analysis – technologies of humility as I have termed them – using methods that look back on the past, with due respect for the multiplicity and ambiguity of human experience. From historical experiences of risk in the real world, including the Asian tsunami and Hurricane Katrina, four analytic focal points have emerged as salient for risk governance: framing, vulnerability, distribution, and deliberative learning. These draw attention to the social and political environments in which risk morphs into reality and shift the weight of analysis from causes to consequences. They not only organize thought in new ways but hold out the promise of wider political engagement in the practices of risk analysis. In these respects, the technologies of humility may provide a bridge to a post-enlightenment politics that is not afraid to confront its unruly past and its ungovernable contradictions.

Notes

1. See, for instance, the argument that genetic explanations artificially narrow the problems of biological development in Richard Lewontin, *The Triple Helix: Gene, Organism, and Environment* (Cambridge, MA: Harvard University Press, 2000).
2. Ian Hacking, *The Taming of Chance* (Cambridge: Cambridge University Press, 1990). See also Theodore M. Porter, *The Rise of Statistical Thinking 1820–1990* (Princeton, NJ: Princeton University Press, 1986).
3. US Department of Energy, *The Manhattan Project: An Interactive History*, http://www.mbe.doe.gov/me70/manhattan/trinity.htm.
4. For an earlier exposition of this concept, see Sheila Jasanoff, 'Technologies of Humility: Citizen Participation in Governing Science', *Minerva* 41: 223–244 (2003).
5. On the role of expert advisory committees in dealing with some of these issues, see Sheila Jasanoff, *The Fifth Branch: Science Advisers as Policymakers* (Cambridge, MA: Harvard University Press, 1990); on differences in national approaches to assessing and managing risk, see Ronald Brickman, Sheila Jasanoff and Thomas Ilgen, *Controlling Chemicals: The Politics of Regulation in Europe and the US* (Ithaca, NY: Cornell University Press, 1985) and Sheila Jasanoff, *Designs on Nature: Science and Democracy in Europe and the United States* (Princeton, NJ: Princeton University Press, 2005).
6. Ulrich Beck, *Risk Society: Towards a New Modernity* (London: Sage, 1992).
7. See, for example, David Harvey, 'The Environment of Justice', in Frank Fischer and Maarten Hajer, *Living with Nature: Environmental Politics as Cultural Discourse* (Oxford: Oxford University Press, 1999), pp. 153–160.
8. Langdon Winner, 'On Not Hitting the Tar-Baby', in *The Whale and the Reactor: A Search for Limits in an Age of High Technology* (Chicago: University of Chicago Press, 1986), pp. 138–154. Also Sheila Jasanoff, 'Civilization and Madness: The Great BSE Scare of 1996', *Public Understanding of Science* 6: 221–232 (1997).
9. Alvin Weinberg, 'Science and Trans-Science', *Minerva* 10: 209–222 (1972).
10. The most authoritative exposition of these views for policy purposes was a report of the US National Research Council. The report laid out the classic risk assessment-risk management paradigm, which holds that the scientific task of assessing risk should be separated as far as possible from the social and political tasks of managing risk, See NRC, *Risk Assessment in the Federal Government: Managing the Process* (Washington, DC: National Academies Press, 1983).
11. For an elaboration of this position, see John D. Graham and Jonathan B. Wiener, *Risk vs. Risk: Tradeoffs in Protecting Health and the Environment* (Cambridge, MA: Harvard University Press, 1995).
12. Influential articulations include Cass Sunstein, *Risk and Reason* (Cambridge: Cambridge University Press, 2002); and Stephen Breyer, *Breaking the Vicious*

Circle: Toward Effective Risk Regulation (Cambridge, MA: Harvard University Press, 1993).

13 See, for example, Paul Slovic, Baruch Fischhoff and Sarah Lichtenstein, 'Facts vs. Fears: Understanding Perceived Risk', in Daniel Kahneman, Paul Slovic and Amos Tversky (eds.), *Judgment under Uncertainty: Heuristics and Biases* (New York: Cambridge University Press, 1982), pp. 463–489.

14 See, for instance, Michael Power and Bridget Hutter (eds.), *Organizational Encounters with Risk* (Cambridge: Cambridge University Press, 2005).

15 Charles Perrow, *Normal Accidents: Living with High Risk Technologies* (New York: Basic Books, 1984).

16 Diane Vaughan, *The Challenger Launch Decision: Risky Technology, Culture, and Deviance at NASA* (Chicago: Chicago University Press, 1996).

17 As of 15 September 2005, Reuters Tsunami Aidwatch reported that donor governments had pledged over $7 billion and private donors some $5 billion in aid for emergency relief and reconstruction. Reuters AlertNet, 'Big Tsunami Donors Allocate Three-Quarters of Funds', http://www.alertnet.org/thefacts/aidtracker/ (accessed September 2005).

18 Oxfam Briefing Note, 'The Tsunami's Impact on Women', March 2005.

19 I was in Chennai, India, on the day of the disaster and remained in India for the following two weeks and so had the opportunity to follow the media coverage while it was unfolding.

20 Oxfam, 'Poorest People Suffered the Most from the Tsunami', Press briefing, 25 June 2005, http://www.oxfam.org/eng/pr050625_tsunami.htm. Oxfam cites three reasons: greater vulnerability; coincidence (impact was greatest in poor regions); and focus of some relief efforts on landowners and businesses.

21 Jennifer Medina, 'In New Orleans, the Trashman Will Have to Move Mountains', *New York Times*, 16 October 2005, p. A1.

22 Adam Nossiter, 'Thousands of Demolitions Are Likely in New Orleans', *New York Times*, p. A1, 23 October 2005.

23 'Barbara Bush: Evacuees Doing Very Well', *Daily Telegraph*, 5 September 2005, http://www.telegraph.co.uk/news/main.jhtml?xml=/news/2005/09/07/ubush.xml&sSheet=/portal/2005/09/07/ixportaltop.html.

24 On the destruction of solidarity, see Kai Erikson, *Everything in its Path: Destruction of Community in the Buffalo Creek Flood* (New York: Simon and Schuster, 1976).

25 Jasanoff, 'Technologies of Humility'.

26 Donald A. Schon and Martin Rein, *Frame/Reflection: Toward the Resolution of Intractable Policy Controversies* (New York: Basic Books, 1994). See also Jasanoff, *Designs on Nature*.

27 Paul C. Stern and Harvey V. Fineberg (eds.), *Understanding Risk: Informing Decisions in a Democratic Society* (Washington, DC: National Academy Press, 1996).

28 For some examples, see Alan Irwin and Brian Wynne (eds.), *Misunderstanding Science? The Public Reconstruction of Science and Technology* (Cambridge: Cambridge University Press, 1996).

29 David Ellwood, 'Empowering the Poor', *Boston Globe*, 27 September 2005, http://www.ksg.harvard.edu/ksgnews/Features/opeds/092705_ellwood.htm.

3
Culture, Structure and Risk
Charles Perrow

The Precautionary Principle is not going to do us much good; nor is the notion of a Culture of Prevention. They are unwitting evasions of the role of power in society. The Precautionary Principle says 'Be aware', think of the consequences of what you plan to do. Of course, this is sound advice, but it can be applied by both sides in any debate or conflict, and neither side would say we should act without thinking of the consequences. Those who oppose the Kyoto Treaty which would curtail the use of fossil fuels exercise the Precautionary Principle when they point to what they say would be the suffering of countless poor people if industrialization slows down, as it surely would have to. Both sides of the controversy over genetically modified foods look ahead to the consequences of allowing it as well as banning it. No one is against being aware of possible consequences; so the principle, by itself, will not settle any debates.

Nor can a plea to develop a Culture of Prevention be much more than a 'motherhood' item (because everyone is for motherhood). Where do cultures of prevention or risk cultures or safety cultures come from? Certainly not from pronouncements. Cultures are 'dependent variables', the results of other forces, developed to explain and legitimate practices, to provide ways of thinking and seeing that are compatible with present existence and experience. The 'independent variables' are many, and include climate, topography, natural resources, science and technology, human cognition and demographic patterns as they are shaped by diseases, natural disasters and diet. These shape life-chances, and above all, the structure of social groups and societies. Culture flows from social structure, not the other way around. Culture is the 'sense-making' activity that 'explains' and reproduces the structure.

The most important determinant of structure for the modern age, holding constant, or even ignoring, the independent variables I have just listed, is economic and political power. This basically accounts for culture. (It is more complicated than that, of course; once established, a culture can resist to some degree attempts by the powerful to change economic and political variables, but rarely can culture change them.) In the West, these powers favour hierarchical organizations, big organizations and organizations only weakly regulated by central governments. There is no room for a significant Culture of Prevention under this social structure, though there will always be small skirmishes and even some victories.

For example, it was not inevitable that the West became dominated by giant economic, for-profit, largely unregulated corporations. In a book on the origins of corporate capitalism in the United States, I tried to demonstrate that alternative paths of development were available in nineteenth-century America, some vigorously put forth by citizens and politicians (Perrow 2002). The economic and political structures were ill formed at the time, and consequently the dependent variable, the culture, was sufficiently open to provide justifications for a variety of paths. The one the US settled on differed from that of other industrializing nations and came about because of a series of almost accidental events and timing, and because of a state that was left weak because of the immigrants' experience with strong states in England and continental Europe. Private interests, unchecked by the state, moved in and developed a capitalist economy, which then gradually created a culture that accepted and justified the economic structure.

Economic elites have shaped our culture in ways that make our concerns about risks truncated. We focus on prevention and mitigation, and have to be, because the alternative – reducing vulnerabilities – is incompatible with our economic structure.

Mitigation means responding to damage after the fact. Here we have first responders, such as police and fire and voluntary agencies. Only 8 per cent of the new spending on 'homeland defence' in the 2003 US budget went to mitigation (Rudman et al. 2003). Furthermore, we have a 'panic' model of behaviour, which mistakenly limits information, and this breeds scepticism. Much research shows that the public rarely panics. They fled collapsing buildings during the 9/11 disasters – but not in panic. Indeed, the public stopped and helped those that needed it, and went back to help others. Other research shows that an informed public responds rationally and creatively to disasters (Clarke 2002; Tierney 2003). But with the panic model put forth by the US government and accepted by the press, inventive responses by the public will be suppressed; the officials will take charge. This official model also legitimates the centralization of responses, whereas research shows the most effective response comes when decentralized units are free to act on the basis of first-hand information and familiarity with the setting. Organizational leaders are generally reluctant to give up power for obvious reasons, and these are structural, not cultural reasons.

A second response is limiting the damage. This involves such things as building codes that cover structural standards, protection of hazardous materials and establishing evacuation plans. The US does only moderately well here; too many powerful business interests resist reasonable standards, and politicians are not disposed to enforce standards or to disrupt dense cities with evacuation routes. After floods and hurricanes we routinely find that buildings did not meet the established codes, just as we routinely find that fire exits were locked when there is a fire in an entertainment centre.

A third response is preventing damage. Here we have alarms and warning systems and data encryption in industry, inspections of homes, surveillance of mosques, and border controls to prevent terrorists and weapons of mass destruction from entering the country. This is where most of the US homeland security budget goes. There are great economic opportunities here for business and industry. As one consultant says, any business can find opportunity in the homeland security budget (Harris 2002; Shenon 2003a and b), but the biggest tranches seem to go to businesses involved in detecting hazardous materials (potential weapons of mass destruction) at ports, spying on citizens through searching massive databases collected by business and the government, and identification devices (eye-prints, gait, body type). Even this is not done economically or well. The Inspector General in the new Department of Homeland Security disclosed substantial fraud in the awarding of contracts, including a fraud of $10 million by Boeing, the aircraft manufacturer. He was quickly fired. Subsequent releases from his department disclosed that the department appeared to be intentionally distributing the money as widely as possible instead of concentrating it in the biggest ports or other locations that intelligence reports suggested were most likely to be future targets.

'Major ports like New York, Los Angeles, Long Beach and Oakland received large allocations. But smaller grants went to ports in places like St. Croix in the Virgin Islands, Martha's Vineyard, Mass., Ludington, Mich., and six locations in Arkansas, none of which appeared to meet the grant eligibility requirements,' the audit stated. The largest sums were given to private firms to refurnish such things as lighting and detectors – expenses they could, and should, pay for themselves (Lipton 2005).

In sum, it will take more than a change of culture to rectify the deficiencies in the areas of mitigation, damage limitation and prevention. It will take changes in the power structure.

The most promising response, to reduce vulnerabilities rather than to limit or prevent them, is the least considered, and it is what I will focus on. (I think Europe has done a better job with basic vulnerabilities than the US, but I have not lived long enough in Europe or studied it closely enough to be certain.)

Our vulnerabilities to nature, industrial errors and deliberate acts such as terrorism can be addressed by examining four concentrations in our societies:

1. Concentrations of toxic and explosive substances (so-called 'hazmats'), in storage and process industries, wherever they may be.
2. Toxic and explosive concentrations that are in heavily populated areas.
3. High-density settlements in areas vulnerable to any of the three disaster sources.
4. Concentration of economic power in private organizations that are essential to our critical infrastructure

Take the risk of disasters from severe weather in the US. With rising income inequality since the 1970s, more rich people have moved to warm areas with lovely sea views, and serving them creates jobs that have drawn the less rich to these areas. The correlation between income and population growth in Florida, the Gulf States and coastal southern California is high. The problem is not nature; the frequency of severe weather on the East Coast, and Florida in particular, has actually declined in the last 50 years. (We were at the low point of the hurricane cycle.) But the number of events declared disasters by the US government increased about four times, because as the years went on there were more targets for the storms. The cost of these disasters rose 14 times. There were more people and more valuable properties are at risk.

Even so we have been quite lucky. New Orleans was almost hit twice in the last ten years by force 4 hurricanes (force 5 is the worst), but their paths are quite irregular and they veered off in the last hour. New Orleans is a gem of city in the downtown area, which houses the famous French Quarter. But it is an empty soup bowl, 10 feet below the surrounding levees. A direct hit by even a force 3 hurricane would fill the bowl, possibly killing 75,000–150,000 according to Red Cross estimates.

We cannot rely on the Precautionary Principle to stem income inequality, government-subsidized storm insurance, local political resistance to expensive building standards or the economic opportunities to builders. All we do is provide modest increases in building standards and better emergency warnings, and even here the rich can get out of harm's way more easily than the poor that serve them. The Netherlands is much more at risk from severe weather, but its greater income equality, and a strong government not dominated to the same degree by business interests, means that more effective safeguards are in place.

Or, moving to southern California, economic policies, such as the location of defence industries, the diversion of water to wasteful agricultural practices and to cities built on the barren (but sunny and warm) coastline, and ample cheap migrant labour – all this has meant the area has become vulnerable to rainstorms and mudslides, floods, polluted land and air, wildfires, and made moderate earthquakes more destructive. Nature has not changed. Economic policies that are made possible because of political power have wrought the damage. The culture that grows in southern California, Florida and Louisiana evolves out of these policies; the people moving there did not come with a Risk Culture, any more than those who settle on the flood plains of the great Mississippi and drown. These areas are where the jobs are, as well as the economic opportunities, or if you are already wealthy, the sun in the South and the canyon views of trees and wilderness in California that are now being ravaged by wildfires and mudslides.

It is similar with industrial and technological disasters, which have been mounting in the US in recent decades. It is not a risk culture that puts unskilled workers in the dangerous subcontracted jobs in the oil and chemical industry, or behind the wheels of big trucks where, by obeying the speed limits, you may lose your job because of 'low productivity'. The risk of the worker being fired is much greater than the risk of having an accident, but we solemnly declare that he is not sufficiently 'risk-averse' and say he should change his culture.

Microsoft was not guided by a risky culture when it refused to make its software products secure; it knew that there would be an outcry about its clunky, clumsy programs. (Those produced by Apple and Linux are superior.) But it knew that with its enormous market dominance it would suffer no economic harm, but would save money and thus increase its profits. But it puts all users in our critical infrastructure at greater risk. We have a monoculture with Microsoft, and monocultures are vulnerable (Geer et al. 2002).

It is more complicated, of course, but does not gainsay the conclusion. Microsoft's clunky code, built on layers of legacy codes to keep up with the small competitors snapping at its heels such as WordPerfect and Apple, did little more than inconvenience users when the users were stand-alone home computers and company workstations. When the Internet came along – and Microsoft did not see it coming for some years and again had to play catch-up and use its market power – the machines became linked. The bank's computers had access to all sorts of financial markets, and that made security from hackers and thieves (and now terrorists) a big factor. Partly because it was such a big target – and this chapter is about the vulnerability of big targets – hackers took pains to bring it down and expose it. As the Microsoft Windows system captured 95 per cent of the market, organizations with safety-critical products were forced to use it, regardless of its unreliability. The US Air Force could not start from scratch to build an operating system, so its fighters and bombers fly with Windows. The Air Force had to re-engineer parts of Windows, I am told, to ensure safety, but with the lurking bugs in such a program we cannot be sure. Microsoft software now threatens the security of huge financial transactions, of nuclear plant and industry controls, and defence department data. Not all software failures can be linked to Microsoft, of course, but many can, and Windows-based software is now critical for most of the things that can kill: heart pumps, infusion pumps, pacemakers, medication distributions in hospitals; military weapons and fighting platforms such as aircraft, tanks and ships; and now the safety of our most widely used weapon, the automobile.

In the financial area the risk is not from hackers, but from thieves, businessmen in a way. It is unpublicized, but is estimated to be in the billions of dollars each year. Citizens eventually pay for the theft allowed by Microsoft, and their policies are a result of their market power. In the defence area, the threat is also great. Terrorists currently favour more dramatic human targets, but sophisticated international terrorists have an open invitation to attack our critical infrastructure because of Microsoft's economic hegemony.

With deliberate disasters, such as terrorism, cults and crazies, any Culture of Precaution is largely irrelevant. It is our social structure and economic and political policies that make our defence so ineffectual. We thoughtlessly, sometimes even deliberately I suspect, demean the culture of the Muslim world. And we amass the world's resources while the Third World gets poorer – not only relative to our wealth, but in recent decades, in absolute terms. The US is despised by those groups that threaten us and it is easy to understand why. No Culture of

Precaution – say, changing the culture of our mass media by portraying nice Muslim families on a sit-com – will change that. As with Microsoft, the US government has international market power, but the government's actions, like Microsoft's, make us vulnerable in non-market ways.

The way we have formatted the discussion of risk in our popular culture and academic literature also tends to avoid the issue of power. Everything is connected in our world, we say, and that is true. But the most important thing about the connectedness is often overlooked: most of it involves dependencies rather than interdependencies. The Third World is not interdependent with the First, but dependent on it. Dependency displays power. Interdependency means reciprocity, mutual affect and mutual adjustments. It also means choice, which is achieved through redundancies, where, for example, a firm has many suppliers to choose from and many customers to sell to. If the suppliers also have many firms to sell to, and the customers many firms to buy from, dependencies are reduced. When they are reduced, there is an incentive for mutual adjustment. The supplier suggests a design change to the producing firm that will benefit both, and can do so since neither the supplier nor the producing firm is dependent on the other. The freedom to seek other suppliers or other producers allows each to experiment and innovate more than if either was dependent. The behaviour of both is affected by each other and the diminished power relationship.

This in turn suggests the efficiency of having goods produced by many firms, and hence smaller ones, serviced by many suppliers and sold by many customers – all smaller than would be the case if dependency, rather than interdependency, were the case. Deconcentrating an industry in this fashion, building in redundancies and interdependencies, reduces the scope of potential disasters.

It might be argued that it is also economically inefficient, but I would disagree. With regard to the firms that sit astride our critical infrastructure, the ones whose failures could have catastrophic consequences for our economic or physical well-being, their scale economies are well beyond any production efficiency requirement. They are giants because they seek market control and political influence. There are four examples of systems which are very large, but made up of small, independent units; that are moderately to extremely efficient; and very resilient and reliable. They are the US electric power grid, the world-wide Internet, networks of small firms in a variety of nations and, alas, terrorist networks. They all evidence interdependencies and redundancies.

We will start with size. The Internet is the world's largest system, embracing the globe. Nothing compares to it in terms of size. The US power grid has been called the world's largest machine, reaching from coast to coast and into Mexico and Canada (though made up of three regional systems, they are connected to each other).

Networks of small firms are much smaller of course, but their output can be much larger than that of one or two large companies making the same products. In fact, typically they displace a few large firms. The point is that the small size of most of the firms in the networks is not an economic drawback, but turns out to be a virtue. They are most famously prominent in northern Italy where an industry making machinery, scientific instruments, furniture or textiles and clothing will range from a few dozen firms to hundreds, all interacting.[1] Finally, while reliable information is not available, it is estimated that the al-Qaeda terrorist network is made up of thousands of cells.

Our four examples point to two important size considerations: the systems can expand easily, and can increase in size without increasing their hierarchies, that is, without encumbering themselves with layers of managers and all the associated costs and complexities. Thousands of new users have joined the Internet every day for years. The power grid can add new lines, territories and capacities rather easily; so can networks of small firms and terrorist groups. This is associated with the 'power law' distribution of nodes in these networks. While there is a very tiny number of absolutely essential nodes, the vast majority of nodes have only a few connections to other nodes, so adding them does not affect the vertical structure. But only a few connections are needed to be able to reach the whole vast network of Internet users, power suppliers, small firms or terrorists cells, so efficiency does not decrease with size. Even the criticality of a tiny number of key nodes in the Internet and the power grid is rarely a vulnerability because of extensive redundancies designed into these systems. In all these respects the networks are very different from traditional organizations, such as firms.

Next, consider reliability. The Internet is remarkably reliable, considering its size and what it has to do. Computers crash many more times than the Internet, and Internet crashes are generally very brief (excepting deliberate attacks). The reliability of the US power grid has been very high, with major outages occurring only about once a decade. It is true that there have been very serious blackouts in the US, and abnormal accident theory would say they are to be expected because of interactively complexity and tight coupling (Perrow 1999). But these kinds of accidents are not just rare, as normal accident theory would expect, but must be considered exceedingly rare given the excessive demands on the system and its size and complexity. Much more likely to cause failures are production pressures, forcing the system beyond design limits, and of course deliberate destabilization to manipulate prices, as in the Enron case in California.

Between 1990 and 2000 US demand increased 35 per cent, but capacity only 18 per cent (Amin 2001). One does not need a fancy theory such as normal accident theory to explain large failures under those conditions. (Indeed, one does need a fancy theory, such as a network theory that gives a role to interdependencies and redundancies, to explain why there were so few failures under these conditions.) One failure in 1996 affected eleven US states and two Canadian provinces, at an estimated cost of $1.5–$2 billion. Even more serious was the 2003 north-east blackout. Since the extensive deregulation of the 1990s we can expect more failures as maintenance is cut and production pressures increase. But I am more struck by the high technical reliability of the power grids than by the few serious cascading failures it has had in some 35 years. Without centralized control, despite the production pressures of mounting demand and despite increased density and scope, it muddles through remarkably well, if it is not manipulated by top management and banks.

The reliability of networks of small firms is more difficult to assess, since there is no convenient metric. But students of small firm networks attest to their robustness, even in the face of attempted consolidations by large organizations. Saxenian (1996) effectively contrasts the decline of the non-networked group of high-tech firms around Boston's Route 128 when federal funding declined and Japanese mass-production techniques matured, with the networks of small firms in California's Silicon Valley, which charged forward with new innovations for new markets. Despite predictions of their immanent demise, dating back to the 1980s when they were first discovered and theorized (Harrison 1994), the small firm networks of northern Italy have survived. In the US the highly networked bio-tech firms are prospering despite their linkages with the huge pharmaceutical firms (Powell 1996). Particular firms in small firm networks come and go, but the employees and their skills stay, moving from one firm to another as technologies, products and markets change.

The reliability of terrorist networks also seems quite high. Rounding up one cell does not affect the others; new cells are easily established; the loosely organized al-Qaeda network has survived at least three decades of dedicated international efforts to eradicate it. There are occasional defections and a few penetrations, but the most serious challenge to it has been the lack of a secure territory for training, once the Taliban was defeated in Afghanistan. (Some argue that al-Qaeda *per se* is increasingly just one part of a leaderless network of Islamic terrorists.)

Can huge, decentralized networks of small units be efficient? It appears so. The Internet is incredibly efficient. Power grids have become more so as they add 'intelligent agents' (though the concentration of generation and distribution firms reduces maintenance and thwarts needed expansion of the grid). Small firm networks routinely out-produce large, vertically integrated firms. Network economies of scale replace those of firm size and rely in part on trust and cooperation, allowing strong competitive forces only when the overall health of the network is not endangered. Transaction costs are low and handled informally. Finally, terrorist networks largely live off the land, can remain dormant for years with no maintenance costs and few costs from unused invested capital; they are also expendable.

I have tried to establish that decentralized systems can be reliable and efficient, but does it follow that the systems responsible for our basic vulnerabilities could be organized like this? The Internet already is, though that is changing. The security vulnerability of the Internet, making it open to terrorist attacks, could be greatly reduced, for instance by making providers such as Microsoft liable for having code that is easily 'hacked'. The electric power grids will remain reliable if maintenance and improvements are required, which could be done through legislation and liability. Deconcentrating industries that deal in hazardous materials, such as petroleum and chemical firms, could greatly reduce vulnerabilities there (along with heavy regulation), and the power the firms would lose would be market power. Without market power they will be more sensitive to their accident potential. Research and development, which might need large amounts of capital and might have to be centralized, could be detached from production, storage and delivery, which could be decentralized. Population concentrations in risky areas is not a 'system' that could be decentralized along the lines of our four examples, so in this case the reform must depend on regulations and improvements in the insurance and liabilities area (e.g. stop federal subsidization of disaster insurance; allow federal aid only if federal standards have been met; increased inspection and penalties; etc.). None of this would be easy, but none of it is inconceivable.

Reducing vulnerabilities means, to give a few examples:

- keeping people off of flood plains – use the plains for farming, not for cities with industrial concentrations of dangerous substances;
- allowing forest fires to burn, replenishing the natural cycle, instead of putting them out because people want to live among trees, but won't accept the risk of doing so;

- storing 'hazmats' in remote areas and in smaller quantities, not in huge tanks along Interstate highways in densely populated corridors, where they are vulnerable to floods, bad management and terrorists;
- limiting travel from those areas that have practices that foster dangerous diseases that can become epidemic – they are fortunately small areas, but nobody should leave them without intense medical scrutiny.

These would limit the targets of weather disturbances and other natural threats, of industrial and technological failures, and actions from deliberate agents, such as hackers, crazies and terrorists.

It would not hurt, of course, to develop a foreign policy that diverts much of US defence spending to helping Third World countries. We won't eliminate terrorism, but this would weaken terrorist organizations.

The prospects for achieving all these deconcentrations are, of course, bleak, and growing bleaker as the current administration in the US works through its second term. But we should be reminded that many of them are fairly recent, a matter of two or three decades of political development in the US. In every case I have mentioned of vulnerabilities we have had laws and regulations that address these issues, or we still have them but they are weak, or are not enforced. To take a small example, the government has tried to withhold disaster relief from those who failed to take out subsidized insurance or failed to conform to regulations, but it has backed off when the flood or hurricane actually came. This could be corrected. More important, we have precedents for deconcentrating organizations; thirty years ago we used to have effective antitrust legislation; it could be reinstated. We tried to break up Microsoft; we could try again. We could once again regulate the Internet as a common carrier, allowing free access, instead of declaring it an 'information system', as we recently did, and thus allowing large telecommunication and media organizations to reduce the number of servers and restrict access (Krugman 2002; Levy 2004).

These are all structural, not cultural issues.

Note

1 There is a vast literature on this. Look at the classic M. Piore and C. Sable, *The Second Industrial Divide* (New York, Basic Books, 1984) which gave the notion, discovered by Italian demographers, its first prominence. It is developed in M. Lazerson 'Organizational Growth of Small Firms: An Outcome of Markets and Hierarchies?' *American Sociological Review* 53 (1988) 30–42. For a synthetic overview, see C. Perrow, 'Small Firm Networks', in N. Nohria and R. G. Eccles,

Networks and Organizations (Boston, MA: Harvard Business School Press, 1992): 445–70. For more recent extensions, see A. Amin, 'Industrial Districts', in E. Sheppard and T. J. Barnes, *A Companion to Economic Geography* (Oxford: Blackwell, 2000); and M. Lazerson and G. Lorenzoni, 'The Networks that Feed Industrial Districts: A Return to the Italian Source', *Industrial and Corporate Change* 8 (1999): 235–266.

References

Amin, M. (2001) Toward Self-healing Energy Infrastructure Systems, *IEEE Computer Applications in Power*, January: 20–28.

Clarke, L. (2002) Panic: Myth or Reality? *Contexts* 1(3): 21–7.

Geer, D. et al. (2002) *Cyber Insecurity: the Cost of Monopoly.* http://www.ccianet.org/papers/cyberinsecurity.pdf, CCIA.

Harris, S. (2002) *Tech Insider: The Homeland Security Market Boom*, Govexec.com. 2005.

Harrison, B. (1994) *Lean and Mean: The Changing Landscape of Corporate Power in the Age of Flexibility* (New York, Basic Books).

Krugman, P. (2002) Digital Robber Barons? *The New York Times*.

Lazerson, M. (1988) Organizational Growth of Small Firms: An Outcome of Markets and Hierarchies? *American Sociological Review* 53: 330–342.

Lazerson, M. and G. Lorenzoni (1999) The Networks that Feed Industrial Districts: A Return to the Italian Source, *Industrial and Corporate Change* 8: 235–266.

Levy, S. (2004) Net of Control, *Newsweek*.

Lipton, E. (2005) Audit Faults U.S. for its Spending on Port Defense, *The New York Times*: A1.

Perrow, C. (1992) Small Firm Networks, in N. Nohria and R. G. Eccles, *Networks and Organizations* (Boston, MA: Harvard Business School Press): 445–470.

Perrow, C. (1999) *Normal Accidents: Living with High Risk Technologies* (Princeton, NJ: Princeton University Press).

Perrow, C. (2002) *Organizing America: Wealth, Power, and the Origins of Corporate Capitalism* (Princeton, NJ: Princeton University Press).

Piore, M. and C. Sable (1984) *The Second Industrial Divide* (New York: Basic Books).

Powell, W. W. (1996) Inter-organizational Collaboration in the Biotechnology Industry, *Journal of Institutional and Theoretical Economics* 152: 197–215.

Rudman, W. B. et al. (2003) *Emergency Responders: Drastically Underfunded, Dangerously Unprepared* (Washington, DC: Council on Foreign Relations).

Saxenian, A. (1996) *Regional Advantage Culture and Competition in Silicon Valley and Route 128* (Cambridge, MA: Harvard University Press): xi, 226 p.

Shenon, P. (2003a) Antiterror Money Stalls in Congress, *New York Times*: 1, A21.

Shenon, P. (2003b) Former Domestic Security Aides Make a Quick Switch to Lobbying, *New York Times*: 1, A20.

Tierney, K. (2003) Disaster Beliefs and Institutional Interests: Recycling Disaster Myths in the Aftermath of 9/11, in L. Clarke, *Terrorism and Disaster: New Threats, New Ideas. Research in Social Problems and Public Policy* (New York, Elsevier Science): 33–51.

4
The Principle of Precaution in EU Environmental Policy
Bärbel Höhn

The global risk society

Social wealth is connected systemically with major societal risks. This is now the norm. In the past 20–30 years, our technological ability to exploit natural resources has assumed a completely different dimension and now affects all areas of life. A new, unprecedented situation has arisen. The great risks – from the dangers of nuclear energy to the greenhouse effect – are no longer restricted in time or space as they used to be. Their impact is now global. We have no idea how long they might last, nor is there any way we can calculate their repercussions or compensate for them. Nobody can escape the potential consequences. To overstate the case: in terms of their impact, the great risks are now distributed 'democratically' across the globe: every single country is in danger.

With these developments in mind, experts and politicians in Europe now speak of a global risk society and formulate this in far more certain terms than their counterparts in the United States. Confronted with these global risks, conventional security considerations appear fragile, for there are no plans for providing aftercare should the worst-case scenario become a reality. In our risk society we have no choice but to provide precautionary care. Having said that, however, this is no reason for us, in Europe, to abandon our passion for invention or stop embarking on voyages of scientific and technological discovery. In contrast to the United States, where many people unswervingly place their trust in scientific and technological progress, Europeans are giving greater thought than ever before to nature and the possible consequences of science and technology. This shift in attitude has brought about some significant changes in EU policies. For in Europe, the risk society has evolved as the dark side of the history of technological progress.

One hundred and fifty years ago, the German philosopher Karl Marx saw the fundamental problem of modern capitalism in a potentially unrestrained free market, which inevitably generated great economic and financial crises. One need not be a Marxist to see the relevance of his forecast to our present period. I cannot help thinking of the French marine researcher and ecologist Yves Cousteau, who remarked that the natural environment suffered considerably under the free market, where everything had a price yet nothing was of any value. How are we supposed to justify environmental and natural protection on the basis of profitability criteria?

The global pressure to maximize profit, gain market advantage and achieve economic dominance is creating enormous environmental risks across the world. In this context, we must also speak of a global class society if we want to examine the real environmental risks more closely. Twenty per cent of the world's population consumes approximately 80 per cent of the world's raw materials and resources. Twenty per cent of the world's population is responsible for 80 per cent of global air pollution and emissions. Those who make the smallest contribution to global environmental risks have to bear the brunt. It is they who suffer most from the environmental catastrophes caused by climate change and face the severest consequences of the global water shortage and pollution of the oceans. Even in the richest countries of the North the old class characteristics of the 1950s and 1960s are becoming ever more pronounced. Our societies are not only being increasingly stratified along the lines of rich and poor, educated and uneducated, but also – following the criteria of environmental medicine – divided into the healthy and the sick. Environmental diseases are not only indicative of social status, but also demonstrate that in Germany (for example) people generally have to bear the burden of the complex socially conditioned environmental risks individually, i.e. in the private sphere. In our society unfortunately, there is a widespread tendency to privatize risks. Political strategies must therefore be designed to ensure unconditional and rigorous intervention to protect the interests of both the individual and the common good.

The culture of precaution

The notion of a culture of precaution means far more than politics. It means, first and foremost, sound common sense, which we apply without thinking in everyday life and is encapsulated in phrases such as 'prevention is better than cure' and 'better safe than sorry'. We weigh up the consequences of our actions: 'Take care,' the Americans say. Who would think of getting into a car and driving off without first putting on a safety belt: 'Show a little restraint, buckle up.' Who feels good about eating something they know is going to seriously harm their health?

We always behave very responsibly towards ourselves when we are fully aware of the negative consequences of our actions. We act with special care when we have to bear the consequences immediately. This is very true, for instance, when it comes to the food we eat. When consumers go shopping, they consider very carefully whether they are willing to buy genetically modified products. However, before they can make such a decision, they need to be well informed. They must be given a chance to take precautionary measures to protect themselves.

This is where political intervention is important. Precautionary state policies make it possible to take into consideration all those things that human beings and nature depend on for life: things that are often neither articulated nor translated into practical measures when acute conflicts of interest arise. The protection of individuals from environmental hazards, the protection and preservation of the natural bases of life (as stated in the German Basic Law and in the Treaty of Amsterdam) are important to me as an environmental politician: I am committed to serving the interests of the common good. If precautionary politics cannot mobilize resistance against commercialization, what can? The state and politicians must offer resistance on behalf of society in trust, as it were, for all citizens, male and female.

The 'precautionary principle' follows a tradition in European politics that is committed to serving the common good and sees state regulatory policy as a corrective to the forces of the free market. This new sensitivity towards sustainable development is reflected in the European concept of risk prevention, which was anchored in the Maastricht Treaty of 1992. In November 2002, the European Commission expressly agreed to apply the precautionary principle in the monitoring of scientific and technical innovation and the introduction of new products.

The corresponding communication issued by the EU Commission on the precautionary principle states that any application submitted for an experiment, a technical process or for the introduction of a product must be examined and, if necessary, suspended, if it turns out that: 'there is good cause for concern, on the basis of a provisional scientific assessment of the risk, that the potential dangers to the environment and the health of human beings, animals and plants might be incompatible with the high degree of protection envisaged by the EU.'[1] During this period, the EU has changed its political priorities, shifting from a willingness to take risks to a resolve to prevent them. The EU is the first governing institution in history to have placed human responsibility for the global environment at the centre of its policies. Since then, the EU's policies have been increasingly geared towards taking measures to prevent environmental damage in advance.

The precautionary principle and global politics

We are all aware that in the US 'precaution' and sustainability currently have very little chance of being adopted at the national level. However, what many people in Europe do not realize is that a broad discussion is currently taking place on the 'precautionary principle' in individual states. Many people hold a responsible position on this question in the world's leading industrialized country. I should like to recall the fact that San Francisco is the first city in the US to act in accordance with the precautionary principle. Not only that, Senators John McCain (R-AZ) and Joseph Liebermann's (D-CT) Climate Stewardship Act of 2003 offered the US an alternative way of responding to climate change with broad-based and flexible cooperation. Following the blackouts of 2001, California managed to solve the acute power crisis by pursuing an electricity-saving programme that cut energy consumption by about 20 per cent. California consciously chose *not* to adopt George W. Bush's plan to build hundreds of new power stations.

Europe is not alone in its approach. The precautionary principle has long since found its way into international statements, the most famous being the Rio Declaration of 1992, which was the first of its kind to stress the right to sustainable development. It recognizes the precautionary and polluter pays principles as guidelines. According to Principle 15 of the Declaration: 'In order to protect the environment, the precautionary approach shall be widely applied by States according to their capabilities. Where there are threats of serious or irreversible damage, lack of full scientific certainty shall not be used as a reason for postponing cost-effective measures to prevent environmental degradation.'

Since Rio, the precautionary approach has become a part of international politics and discourse. Unfortunately, however, turbulent developments in the world economy have interrupted the process that started there, triggering stagnation if not retrogression in global environmental and development policy. Even so, the UN Earth Summit held in Johannesburg in 2002 resisted all attempts to give World Trade Organization (WTO) agreements priority over multilateral environmental agreements. The political declaration issued in Johannesburg stated that international regulations had to be established in the near future which would bind multinational groups to precautionary environmental and social standards.

The creation of a world organization for sustainable development is now on the political agenda. An international organization of this kind would be able to link the supranational agreements and programmes on environmental protection with those on combating poverty in the developing countries. It could help ensure that the rich industrialized countries take greater care of their environment and resources, and that they share their technological advantage in the field of precautionary environmental protection with countries in the South.

Unfortunately, the United Nations Environment Programme (UNEP), headed by Klaus Töpfer, has not yet been upgraded to the status of an international UN organization for development and the environment. The German proposal (supported by the EU) to establish a UN international commission on 'sustainability and globalization' as part of a post-Johannesburg agenda has not been followed up either. Yet we urgently need agreements to upgrade global precautionary environmental protection. We need a new approach to international politics, one that allows us to challenge national claims to sovereignty over issues that affect both collective resources and goods belonging to the international community. Climate protection and the diversity of species, for example, are vital to the survival of the human race. As such they must be secured by international agreements. This is a legitimate move under international law.

The global market urgently needs good governance: a world domestic policy without a world government. The vision of sustainability needs counter-elites that oppose the dominance of transnational capital markets; it needs to find counterweights to financial flows and production structures that are increasingly developing lives of their own at a global level. This means good governance, in other words: a political architecture founded not on a world order based one-sidedly on states but on a network of states and international institutions, which also includes civil players from the areas of science and business, as well as from among the NGOs.

Climate policy and precaution

Anyone who wants to gain a more precise idea of how far international political developments have moved towards adopting the principle of 'precaution' should take a look at global energy and climate policy. The energy problem is now global in every respect and encompasses energy shortages as well as the environmental risks related to the consumption of fossil fuels. Climate change is by far the greatest global environmental risk: we are already witnessing climate change, as the dramatic rise in incidences of extreme weather shows. What humanity can hope to achieve at the moment is to prevent sudden catastrophes in the future. We must do everything we can to halt the rise in global temperature and ensure that the temperature is reduced and kept at least 2 degrees below the twentieth century's level. That is the target set by climatologists.

Climate policy thus means, first and foremost, global precautionary environmental protection. It means protecting millions of human beings. The Kyoto Protocol is of paramount importance here. In 1997, the majority of states agreed to ratify the treaty within the framework of a UN climate conference. In the Kyoto Protocol, the industrialized countries promised to cut their emissions of greenhouse gases to at least 5 per cent below the 1990 level. The EU's climate resolutions, which aim at 15 per cent, go way beyond this figure. Germany has committed itself to a 21 per cent reduction. So far it has achieved 18 per cent.

The Kyoto Protocol centres on international trade in emission rights, which will make it easier to protect the environment. Any country that produces in a relatively clean manner can sell certificates. One that generates more pollution has to buy credits. This approach gives industrialized countries an incentive to invest more in both environmental protection and renewable energy in the developing countries, for investors from the North then receive credits for reductions in greenhouse gas emissions. It is easy to imagine how attractive this would be for many investors and how much it would boost innovation in the poorer countries. At the Earth Summit in Johannesburg in 2002, approximately 80 countries declared that they were not prepared to wait until all the countries had signed up to Kyoto. They agreed to do all they could to ensure that a billion people would be supplied with electricity and heating from renewable energy sources by 2015.

In June 2003, EU President Romano Prodi gave a speech at an EU conference on hydrogen and fuel cell technology in which he stated the direction in which the European Union would go: 'It is our declared goal of achieving a step-by-step shift towards a fully integrated hydrogen economy based on renewable energy sources by the middle of the century.' The key question for Europe, according to Prodi, was 'whether we have enough air, land and sea to dispose of the gaseous, liquid and solid wastes from spent fossil and nuclear fuels used to produce energy. The answer is a clear "No". ... The rational solution would be to turn resolutely towards renewable energies ...'[2]

As I have said, in the US 'precaution' and sustainability have very little chance of success. President Bush bases his argument for rejecting the Kyoto Protocol on the fact that climate protection entails economic disadvantages for a highly industrialized country, which are quite unacceptable. To understand this it is important to know that Bush has close links with the US oil and coal industries. Yet his argument, which seemingly makes such good economic sense, is politically devoid of reason. The industrialized countries consume three-quarters of the world's energy reserves and pump three-quarters of the noxious greenhouse gases into the atmosphere. Given this state of affairs, it is our political duty to prevent a climate catastrophe, which could turn out to be far worse in the southern than in our northern hemisphere. One thing is certain when it comes to 'new energy for the world': only if the North succeeds in linking energy supply and precautionary climate protection will the poor countries of the South be able to adopt a new energy policy.

Global precautionary protection through eco-efficiency

Global trends show quite clearly that environmental destruction and the degradation of non-renewable energy resources have increased exponentially during the past few decades. It has long since ceased to be a question of ethically motivated measures to protect the environment and nature. It also is a question of the exploding costs caused by the global exploitation of nature. Ever more money must be raised to repair environmental damage, to remedy the consequences of environmental catastrophes and (what is becoming increasingly important) to exploit new natural reserves. The share of the social product and of private wealth accounted for by environmental and resource costs is rising rapidly. We can less and less afford to place our trust in environmental aftercare alone and simply wait and see what happens.

Consensus prevails among economists and environmentalists that there is an overall decline in the volume of material flows. This development calls for measures, based on the precautionary principle, to protect natural resources: economical use of materials and energies, the recycling of used materials, avoiding the production of waste and waste water, and reducing emissions. The focus must be on resource-conserving production methods, technologies and products. The UN believes that the productivity of energy and resources may be about to rise by the factor of 4 in the industrialized countries. International company networks swear by the shift to ecological efficiency. Product life-cycles – from the initial raw materials to waste disposal – are under scrutiny. The concept of eco-efficiency – together with the model of sustainable development – has meanwhile become part of our global vocabulary. We now talk of a new kind of 'lean production' in the sense of minimized use of materials and energy in a variety of areas including information technology, transport and energy production, football stadiums, and kitchen appliances.

In this connection I should like to say a few words about two scientists, Ernst Ulrich von Weizsäcker and Friedrich Schmidt-Bleek, who work at the Wuppertal Institute for Climate, Environment and Energy. Both scientists have developed models of so-called eco-efficiency that are of regional and global interest. The simple formula underlying both is: apply engineering to reduce the consumption of raw materials.

I shall never forget the famous phrase with which Professor Weizsäcker reduced the factor 4 concept to a simple formula: 'We can produce four times as much wealth with one barrel of oil as we've done in the past.' The Wuppertal Institute prepared a now famous study, *Zukunftsfähiges Deutschland* (Sustainable Germany). This impressively shows that the good life does not depend on a constant increase in the flow of materials and a steady rise in energy consumption. It shows possible alternatives: eco-efficiency and sustainable management. We, in the rich countries, now have to create wealth and establish justice with fewer raw materials and less energy. That is the meaning of a global precautionary approach and, above all, a precautionary policy for the poor countries.

I should like to return to the beginning of my argument. A dramatic change is currently taking place: from a willingness to take risks to the need to implement risk prevention measures. The EU is the first governing institution in history to place human responsibility for the global environment at the centre of its political programme.

Taking precautions against genetically modified food

This brings me to one of the most delicate issues in contemporary politics: the genetic modification of plants and foodstuffs, which also takes us back to the precautionary principle. Major companies involved in genetic engineering now have to apply to the EU for licences to genetically modify plants. One of these is Syngenta, which wants a licence to produce insect-resistant Bt11-corn. The US has filed a complaint with the WTO against the EU. President Bush insists that it is high time the EU abandons its opposition to genetically modified foodstuffs and opens its markets. Genetic engineering is lucrative for large seed producers.

The EU has a different perspective. Over the past five years it has undertaken extensive analyses and prepared estimates of the risks involved in introducing genetically modified organisms and, as a consequence, genetically modified food too. During this period, there was a *de facto* moratorium on the application of genetic engineering in agriculture. The safety risks had to be scientifically established before new genetically modified plants could be authorized in the EU.

As a result, a new EU regulation on labelling came into effect in November 2003, which had to be implemented by manufacturers by April 2004. It contains strict new rules that are intended to limit potential damage by genetically modified foodstuffs. These rules prescribe a licensing procedure designed to offer greater safety than in the past. Despite some grave shortcomings, the prescribed label will offer consumers enhanced transparency. Under the so-called traceability guideline, it will be necessary to monitor the long-term effects of genetically modified foodstuffs.

A precautionary chemicals policy

The new EU legislation on chemicals is also based on the principle of precautionary environmental policy. Above all, it obliges companies to estimate the risks involved in using dangerous waste materials. With this, the EU is the first political institution in the world to shift the risk to companies and make them responsible for proving that their products are safe.

A few years ago, the European Commission published the basic ideas in a white paper entitled REACH. This was followed, in October 2003, by the draft of a new EU directive. The new regulations must now be passed by the European Parliament and the European Council.

REACH (registration, evaluation and authorization of chemicals) is at the heart of the debate. It is a comprehensive piece of legislation on how to proceed with chemicals. The new system was introduced to replace existing regulations and is designed to reduce the toxic effect of chemicals on both the environment and the health of human beings and animals. REACH is also designed to prevent environmental damage before it occurs. Chemicals companies have to assume responsibility for risks by undertaking safety and environmental tests. They must prove that the materials they produce are safe. Shifting the responsibility for the risk onto the companies represents a new chapter in the history of environmental and health risks.

The main goal of REACH was to solve the problem of old substances. To this end, a new safety system was adopted. What this means in practice becomes apparent when one looks at the old system. Nobody in Europe now disputes that there are far too many chemicals that we know far too little about. REACH would be unnecessary if we knew enough about the environmental effects of the chemicals we are using, if we knew precisely what health risks certain chemicals entailed and what we could do to avoid these risks, and whether the old regulations were at least suitable for gradually reducing health and environmental risks. Past experience indicated, however, that the old Toxic Substances Control Act was not sufficient for dealing with problems quickly and, above all, with any degree of permanence.

The so-called system for registering new substances has now come into force and functions quite well. This system applies to all chemicals marketed for the first time since 1981. The real problem, however, lies with the old substances, since these have not been registered. In Europe, most chemical substances are waste materials. These are substances that have been on the market since 1981, the cut-off year. A comparative figure clearly illustrates the nature of the problem: since 1981 some 3,800 new substances have been registered in the EU. However, the number of old substances has been put at around 100,000. It is estimated that approximately 30,000 of these are still being manufactured in significant quantities.

The system for registering new substances does not apply to these 30,000 substances. Not only do we know far too little about them, but our lack of knowledge gives cause for alarm. It is important to bear in mind that most of the chemicals in Europe have not been adequately tested for their effects on our health or the environment. In the past, such gaps in our knowledge have resulted in serious environmental problems. I should like to cite a few examples:

- DDT: A substance that was considered an excellent insecticide until it was discovered that it not only presents a danger to insects, but to other forms of life too.
- PCB (polychlorinated biphenyl): these substances have excellent performance properties as softeners and as condensator oil. It has been established, however, that children who come into contact with PCB may subsequently experience developmental defects.
- CFCs: these substances have been used in many areas as a propellant because they are stable and rarely react in contact with other substances; however, they damage the ozone layer at a considerable height.

The list of hazardous substances could easily be added to. We often speak somewhat cynically of the 'pollutant of the month' when a specific substance catches our attention. We talk of substances which have been produced for years or decades because they have good material or performance properties. However, these substances have not been subjected to precise tests to determine their toxicological and eco-toxicological properties. As a result, a lot of environmental damage was done in the past which cannot now be rectified. In some cases, vast sums of money have had to be invested in cleaning polluted areas. This is the classic case of environmental aftercare, which generally comes too late and is much too expensive. We can no longer allow a situation to arise in which the authorities impose restrictions or ban substances only after the damage has become acute. We cannot afford to have situations in the future in which the burden of proof always lies with the authorities, and that the authorities – and not the polluters – have to prove that a substance is hazardous before any action is taken.

The question of cost always plays a major role in the REACH discussion: 'Isn't this all a bit too expensive for a company?' I should like to present the question of cost differently: What do we do about the huge costs incurred in the past and the present as a result of damage to the environment and health? Who is supposed to meet the cost of treating people suffering from asbestosis? Who is supposed to pay for the removal of PCB from hundreds of buildings?

In the past, there have been countless attempts in the EU to come to grips with the problem of old substances. We have to admit that all such attempts have failed. We now need to rethink. We need a new solution. I am convinced that REACH is a step in the right direction.

Incidentally, the EU estimates that REACH will cost chemicals companies in the EU €7 billion at most (a minimum of €1.4 billion). The European Chemical Industry Council (CEFIC) estimates the costs at between €7 billion and €10 billion. The German Council of Environmental Experts compared the EU's estimate with the annual turnover of the chemicals industry and arrived at the following figures: the additional annual costs incurred due to REACH would have amounted to between 0.06 and 0.12 per cent of their annual turnover for 2001. The fine and special chemicals sector would probably have borne the greater part of the costs. Even there, the additional costs resulting from REACH would have amounted to only 0.2–0.4 per cent of their annual turnover in 2001. These are the calculations of the Council. One thing must be remembered, however: these are averages for all companies computed for a period of 11 years.

A new feature in the REACH system is that so-called downstream users have to fulfil very specific obligations: if they want to use their chemicals in areas not intended by the manufacturer, they must inform the manufacturer accordingly. If they do not, they must perform an exposure estimate for any special use. Importers are subject to exactly the same requirements as manufacturers. They too must register their substances. This can, of course, be difficult for foreign companies. From the European point of view, however, there is no alternative. It is essential that the same requirements apply to all, no matter whether the substances are manufactured at home or have been imported.

The REACH system will replace the present practice of registering and evaluating chemicals. For REACH reverses, as it were, the burden of proof by transferring responsibility to the companies, which now have to assume greater responsibility for their substances. Those that produce and sell a chemical and earn a lot of money from it are expected to assume responsibility for its safety and for any harm it might cause. Furthermore, REACH ensures that downstream users receive better information about the substances they are using.

I am optimistic that REACH will enable us to eradicate the problem of the 'pollutant of the month'. The chemicals companies will obviously profit from this too: for products labelled 'pollutant of the month' always result in a loss of public confidence, which also means a loss of confidence in the company and even the entire sector.

The precautionary principle is intended to prevent damage before it can occur. In the case of technological progress, we must weigh up the benefits against the potentially destructive consequences. I see this as the essence of the precautionary principle. The principle compels us to look beyond the short term. And we must also do this because modern environmental risks always entail the danger that measures will be taken too late, because politicians, the media and the public are far too concerned with the here and now. Today. environmental policy must be conceived on a greater temporal and spatial scale. In other words, it must relate environmental questions to questions about the future.

If we are serious about our future, we need to base our politics on a model of sustainable development. The most important thing of all is that politicians, managers and citizens broaden their horizons. We must grasp the long-term consequences of our actions, as most families do: parents try to ensure that their children have a better future.

Above all, the precautionary principle must be translated into effective politics. For this to happen, politicians need society; they need the broad diversity of initiatives launched by men and women able to articulate their interests. We cannot allow their voices to be drowned out by all the commotion and noise of day-to-day politics.

Notes

1 EU Mitteilung, EU-online, 2 February 2002.
2 For further developments since the 2003 Paris conference, see the official site of the European Hydrogen and Fuel Cell Technology Platform: http://www.HFPeurope.org.

5
The European Social Model from a Risk Management Perspective

Günther Schmid

Introduction

There is much talk today about the 'European social model'; its meaning, however, remains unclear.[1] Most people associate it with generous social security, moderate inequalities in income and corporate governance; some might even add full employment. While many want to maintain these features, others see them as a problem for a sustainable European welfare state. When asked for further clarification and specification, many probably would hint at the alleged opposite: what Europe is not or should not become, namely a neo-liberal model of workfare and tremendous income inequalities as in the United States. People preferring positive and prospective definitions might refer to the social-liberal model, often sold as 'asset welfare capitalism', 'social investment state' or the 'third way' between workfare and welfare.[2]

Although this 'third way' has sympathetic features, it misses one important point in the hidden agenda of the European social model: solidarity with people who do not succeed in the market despite equal opportunities and striving in the competitive game. The objective of this chapter, therefore, is to seek an alternative. The evolution of the traditional welfare state towards an ensuring and enterprising state could be the central organizing idea of the European social model.[3] In operative terms, the ensuring state would marry individual and collective responsibility through the joint management of risks.

The argument will be developed in four steps:

1. What is the changing nature of risks in the labour market? It is assumed that new technologies in information and communication, changes in demography and family structures, the extension of the

common market and competitive pressures from globalization change the nature and incidence of risks in the labour market. New social groups are affected – for instance, young adult women through the increasing compression of their careers, and overall the low-skilled, especially mature adults. The rise of internal risks and the blurring of internal with external risks require a substantial adjustment of the established insurance institutions and a shift from state social policy to joint management of social risks.

2. What are the perceptions and possible reactions of people with respect to these new risks? It will be shown that optimism – that risks can be managed rationally if we only we have the will – is not justified. The psychology of intuitive beliefs and choices points to the bounded rationality of risky choices. People overestimate short-term low risks and underestimate long-term high risks and tend therefore to myopic decision-making or switching between excessive risk-aversion and irrational risk-taking. It will be shown how the institutionalization of opportunity structures can overcome various kinds of asymmetries in risk perception and endorse precautionary steps towards risk management.

3. How should the European social model respond to the employment crisis in Europe? It is assumed that the institutionalization of transitional labour markets, grounded in the concept of shared responsibilities and equality of resources, is an essential part of a proper response. This strategy recommends the extension of unemployment insurance to a work-life insurance consisting of three pillars: state or collective unemployment insurance for pure external risks; mobility accounts and activating labour market policy for the mix of internal and external risks; complemented by various schemes of private insurance for internal risks.

4. Good practices in various European member states will be discussed to show how the theory of social risk management in general and its translation into the concept of transitional labour markets can contribute to the discourse as well as to the evolution of a European social model. Special attention is given to institutional innovations related to work-life insurance supporting people in taking over more responsibility in preventing, mitigating and coping with the new employment risks.

Old and new risks in the labour market[4]

First, there is an increasing low-skill risk. If we take as the main indicator for full employment the European Employment Strategy, namely an employment rate of 70 per cent by 2010, then the breakdown by qualification immediately shows where the main problem lies: high-skilled people surpass the 70 per cent benchmark by about 15 percentage points, and this holds true independently of employment or welfare regime. It is the low-skilled whose opportunities of participating in the labour market are seriously harmed. In Germany, for instance, the employment rate among the low-skilled is 50.9 per cent, more than 30 percentage points below the 83.6 per cent employment level of the highly skilled.[5]

It is important to look also at the other side of the coin, which is unfortunately not well reflected in the European Employment Strategy, to the unemployment rates. While the Lisbon Strategy set the benchmark at halving the unemployment rate to about 5 per cent by 2010, the statistics shows that – with few exceptions – the highly skilled have already reached this level or are even below it, whereas in many countries the 5 per cent benchmark is well out of reach for the low-skilled. In Germany, for instance, the unemployment rate of the low-skilled is, on the average (and apart from huge regional differences), 15.3 per cent against 4.5 per cent of the highly skilled.

The second concern is the increasing number of precarious jobs in the form of fixed-term contracts, temporary agency work or contract work often disguised in the form of self-employment. Why do we find this trend in almost all EU member states? It seems that firms need more flexibility to adjust to an increasingly competitive environment or new technologies. However, job protection is strong in the family-centred employment systems of southern Europe (Italy, Spain, Greece), and is still important in the conservative or corporate employment systems of continental Europe (France, Germany, the Netherlands).

The exceptions to this rule, however, are revealing. It is Denmark and the United Kingdom where the share of fixed-term contracts decreased from an already low level. In both countries protection from dismissal is almost unknown. However, whereas Denmark compensates for the lack of job protection by generous payments in the event of unemployment combined with robust retraining programmes or compulsory work placements, the UK has been able to some extent to stem the tide of precarious jobs with its New Deal programme and successful job-creation machinery.

Of increasing concern is the concentration of precarious jobs among the young. Germany is especially striking in this respect. The burden of risks related to fixed-term contracts lies almost completely on youths (aged 15–25) and young adults (aged 25–35). We know from many studies that fixed-term contracts are often useful entry points to regular work. But unfortunately for many young people (and in some countries the majority) fixed-term contracts are traps into permanently disrupted job careers and often, in the longer term, to social exclusion.[6]

These risks are often aggravated by 'compressed work careers'. This risk denotes the phenomenon of having to fulfil several social roles simultaneously in a short period of working life. It mainly affects young women (aged 20–35). Since labour market participation is becoming the norm for such women, they have to get to grips with at least five social roles more or less at the same time: they have to:

- acquire a good education,
- look for suitable work,
- plan a sustainable career,
- select a suitable partner and
- set up a family at considerable expense for housing and furnishing.

The way work, education and welfare (including the housing market) are organized scarcely helps them to get to grips with these diverse roles. Their transition to a sustainable employment career is seriously jeopardized.

Even if success in this respect does materialize, the associated pressures can lead to physical or psychological stress. A study carried out in the Netherlands has revealed a dramatic increase in the incapacity to work among young women, while an Australian research team speaks of the 'excluded generation'.[7] The relative neglect of this problem for young adults compared to the attention given to older adults is a serious shortcoming in the European Employment Strategy.

This does not mean, of course, that we should abandon attempts to tackle the third new risk: the diminishing earning capacity of mature adults, reflected in employment rates below the full employment benchmark of 50 per cent for most of the EU member states. In Germany, only 39 per cent among the 55–64 age group are employed. A notable exception is Sweden, where 69 per cent of this cohort are actively participating in the labour market.[8] If mature adults become unemployed, they face either a high risk of long-term unemployment or of a drastically reduced income.

All three risks – low skill, job instability and reduced work capacity – have to be considered against the background of the erosion of internal labour markets. From the perspective of risk management, the backbone of internal labour markets was an implicit insurance contract: the employer offered the male breadwinner a family wage, job security and earnings stability over the life-course in exchange for wages below the productivity level at the height of the working career. This implicit insurance contract is breaking down without a clear alternative in sight.

A plausible conclusion would be to extend the principle of insurance to cover – at least to some extent – these new risks.

Before coming to this point, however, it is necessary to look at how people perceive risks and make risky choices in order to have a behavioural background for establishing new institutional arrangements for managing the new social risks.

How people perceive risks and make risky choices

In contrast to the mainstream of rational choice theory the new paradigm of social risk management[9] should take into account that people tend to make myopic decisions, or – depending on the framing of risks – to engage in unreasonable speculations or to extreme risk-aversion. These tendencies are predicted and explained by the theory of intuitive judgements and choice, developed by the Nobel Prize winners for economics, Daniel Kahneman and Amos Tversky (2000). Their prospect and framing theory contains four core ideas:

1. The carriers of utility are events and not states. This means that people assign utility to gains and losses relative to a reference point, which is often the status quo. Critical transitions from one employment status to another during the life-course are prominent events where people compare utility and disutility.
2. In assessing the prospects of events, losses usually loom larger than gains. Many experiments suggest that the loss/gain coefficient is about 2:1, which means that losses are afforded twice the value of gains.
3. There is an endowment effect giving present or past experiences intuitively higher values than future prospects. The German comedian Karl Valentin expressed this when he said: 'In the past, even the future was better.' Thus, the transition from status quo to receiving something is valued less than the transition from status quo to giving up something. This endowment effect often leads to myopic decisions and violates a substantive condition of rationality – utility maximization.
4. Framing may violate another assumption of rational choice theory: the consistency of decisions. Framing theory refers to the fact that alternative formulations of the same situation make different aspects accessible to actual perception, thereby leading to different reactions. The same objective outcomes can be evaluated as gains or losses, depending on the framing of the reference state. In experimental situations, for example, the prisoner's dilemma was played out several times, once

formulated as a Wall Street game and again as a community game. People were more inclined to cooperate in the community game than in the Wall Street game.

These asymmetries of risk perception lead to three hypotheses as crucial starting points for social risk management:

1. If there is a choice between certain and uncertain gains, most people tend to be risk-averse. That is, they choose the certain alternative, even if the objective value of the prospective gain is greater than the value of the certain gain. In other words, they prefer the bird in the hand to the two in the bush.
2. If people have to make a choice between certain and uncertain losses, they tend to be speculative risk-takers. Since they don't like losing something, they tend to prefer the uncertain loss to the certain loss, even if the objective value of the prospective loss is greater than the immediate and certain loss. In other words, they tend to behave like a roulette player who tries to recoup his losses again and again until he ends up with nothing.
3. Most people overestimate minor risks that are immediately in sight, such as the possibility of becoming ill shortly before a planned journey. On the other hand, most people underestimate major risks that lie in the more distant future, such as becoming unable to work through disability or being forced to switch career. In consequence, many people are willing to buy a travel insurance but reluctant to buy disability insurance. They are also less willing to save for education or training that may be necessary in the future. A recent study on European attitudes to the welfare state, for example, found that the majority of voters under-insure against the risk of unemployment.

These three insights can be used to design and gain acceptance for strategies of social risk management:

- The proper strategy against risk-aversion would be to provide for a large opportunity set in critical events during the life-course. This would reduce the high subjective valuation of smaller but more imminent loss(es) against greater but uncertain gain(s) in the future. It would also increase the probability of possible gains by extending and securing the alternatives in event that the first risky choice failed. The preference for 'a bird in the hand' might change if more bushes were in sight. Establishing such a variable opportunity set with 'stepping stones' or 'bridges' is one of the main objectives of transitional labour markets, to which I will refer later.

- The proper strategy against loss-aversion and unreasonable risk-taking would be to establish disincentives to gambling, for example, through high taxation of speculative gains and of large-scale inheritance, which is not re-invested. In this perspective, job protection can be rational in the sense that it avoids speculative search behaviour on the supply side and irresponsible hire-and-fire policy on the demand side. Both behavioural tendencies lead to high fluctuation costs for the whole economy. However, as empirical studies have shown, job protection makes sense only in combination with activation policies, for example, with the obligation of further training or the obligation of flexibility through taking on different tasks within the protected employment relationship. If not, job protection turns into insider monopoly and corresponding segmentation of the labour market.
- The proper strategy against overestimating short-term (small) risks and underestimating long-term (high) risks would be to extend the expectation horizon for people engaging in risky employment relationships. One way to do this would be to establish social rights and entitlements, for instance the entitlement to continuous training opportunities on the supply side. Another way would be to set up incentives or obligations for preventing or mitigating risks with high and long-lasting harm, for instance incentives for preventive measures (such as training investment), or the requirement to participate in universal private or public insurance schemes (such as invalidity insurance), or the requirement for employers to take positive measures towards disabled people, for instance through workplace adjustment or even quotas for the disabled.

How can these insights, together with other aspects of social risk management,[10] be applied to the design and strengthen the European social model, especially in relation to the European employment strategy?

Social risk management through transitional labour markets

The evolution of the social welfare state to the ensuring and enterprising state could be the central organizing idea of the European social model. In operative terms the ensuring or enterprising state would marry individual responsibility and solidarity through the joint management of social risks. The main suggestion is to consider the social dimension as a problem of 'risk management' and to look especially for institutional arrangements that prevent, mitigate and cope with the risks related to critical events during the work life-course. Elsewhere I have identified five types of such critical events:

1. the transition from school to work or between employment and education;
2. the transition between various employment statuses or the combination of different employment statuses;
3. the transition between unpaid (usually family) work and labour market work;
4. the transitions between unemployment and employment; and
5. the transitions from employment to retirement or between (partial) work disability and full employment (Schmid 2002).

Social risk management through transitional labour markets suggests extending the time-honoured unemployment insurance to a kind of work-life insurance.[11] If such a system were to be established in a systematic way and coordinated properly with the other social protection or social security systems, it would be an effective functional equivalent for rigid employment protection systems. The shift from unemployment insurance to multifunctional work-life insurance would not necessarily imply more public expenditure. In many cases, legal entitlements to negotiate collectively or individually on various transitional arrangements would suffice. It could even reduce the tremendous costs of conventional unemployment benefits through transforming welfare benefits into employability investments. However, the concept of 'activation' based on the philosophy of balancing 'individual rights and responsibilities' cannot substitute completely for the need to redistribute means and ends in favour of the losers of structural change due to bad luck or circumstances for which they cannot be held responsible. And even where people might indeed be responsible, they can find themselves in dire situations that require immediate remedies and help. Social risk management through transitional labour markets would periodically redistribute resources in favour of enlarging the opportunity set for all people, independent of their talents and fortune, out of compassion and solidarity.

It is suggested, therefore, that John Rawls' (1990, 2001) theory of justice, further developed by Ronald Dworkin's (2000) concept of 'equality of resources' and complemented by Armatya Sen's (2001) capability approach, be adopted instead of relying on the utilitarian concept of 'equality of welfare'. Following the typology of risk management derived from this framework, work-life insurance would consist of three pillars:

1. The conventional unemployment insurance pillar would be reduced to the core function of covering income risks of involuntary

unemployment through pure circumstances. Since the income portfolio, especially of middle or high wage-earners, is broadening, the financing basis of such schemes would have to include a redistribution element through extending the contribution base, for example including non-wage income from equity shares or other profits, and through abolishing wage-income ceilings. The corresponding reduction or exemption of pay-roll taxes for low-wage earners would help to create jobs in the low-wage sector and increase work incentives through reduced tax bands. In addition, overall income inequality insurance would help to overcome income volatilities due to technological and economic change, thereby encouraging overall risk-taking and so enhancing productivity.[12]

2. Enhancing equality of opportunity though transfers or funds would be financed through general taxes.[13] Selective active labour market policies and state-subsidized mobility accounts could be used to establish drawing rights (Supiot, 2001) or vouchers to 'insure' income risks related to various kinds of transitions during the life-course. Vouchers or tax credits would enable and empower workers legally and financially to engage in transitions and to manage their own transitions adequately.[14] Another strategy for equalizing resources through *ex ante* redistribution would be a basic income guarantee independent of the work-life course, especially basic pensions. Discontinuity of income streams during the life-course, caused by a mix of individual choices, cultural inheritance or 'accident', would thereby not be punished but even encouraged. Labour market policy would no longer only activate people (so far mostly men) into regular jobs, but also activate people (from now on especially men) into unpaid family work or other kinds of civil work.

3. Various individualized social security, employability or mobility, training or working time accounts. The current unemployment insurance contributions could be divided into two parts, one covering the costs of the first pillar, the classical unemployment insurance (unconditional solidarity); the other covering the costs of the third pillar, the individualized mobility accounts, either legally mandated or collectively bargained (negotiated solidarity). Since established (implicit) private insurances, such as seniority wages (especially in internal labour markets) and status-related wages (especially related to formal educational levels and to family status), are eroding new forms of insurances based on individual responsibility will and must be developed. In addition, new techniques of risk management and adoptions of management techniques from financial markets might extend the possibility of private insurance of labour market risks (Shiller 2003).

Finally, some good practices might be at place to give a flavour of how transitional labour markets would work in reality.

Good practices within the new European employment strategy

Social risk management through transitional labour markets would suggest that employment policy has to be directed away from single interventions towards a strategy of institutionalizing a rich set of opportunity structures. For instance, the opportunity to switch from full- to part-time work without losing income and social security; or the opportunity to switch from waged work to self-employment without losing the accumulated entitlements of unemployment insurance; or the opportunity to combine work with training or education without losing the job. This is what transitional labour markets are about. Transitional labour markets intend not only to 'make work pay', but also to 'make transitions pay'.[15]

From the tendency of overestimating short-term small risks and underestimating long-term high risks, it can reasonably be assumed that individuals perceive the risk of being stuck in the low-wage sector to be greater than the risk of long-term unemployment resulting possibly from being too choosy about the jobs they will accept. Active labour market policies, therefore, should not be confined to offering jobs and placing individuals in work. Follow-up measures are required for transforming workfare measures into stepping-stones to a sustainable career.

Conversely, the barriers to taking on a new risky job should not be underestimated. Such barriers may arise from the fact that acceptance of risky jobs means abandoning certainties, even though they may have a lower value than the new employment prospects. These 'certainties' may be of various kinds. The reliability of social assistance benefits possibly supplemented by some work in the black economy is one example; the confidence in one's own productive capacities is another. Taking on a risky new job, however, brings with it the fear of losing these capacities. Daniel Bernoulli, one of the founders of probability theory and thus of risk management, provides an example: even a beggar will not give up begging for a workfare job since he would lose his option to beg.[16] You have to offer him a bit more.

What might this 'more' be? The concept of TLM suggests a whole set of possibilities that are already good practice in some countries or can be added to existing policies, for instance:

- The opportunity to try several jobs without withdrawing benefits immediately if one option does not lead to success, a rule that excludes rigid workfare strategies that do not allow trial and error as a productive job search strategy;
- The opportunity of building up flexible entitlements, for instance the possibility of transforming entitlements to unemployment benefits into entitlements for co-financing continuous training or education on and off the job;
- The opportunity of in-work benefits under certain circumstances, for example if earning capacities are reduced due to family or other care obligations.

Second, unemployment insurance in place provides inadequate security in particular for the new labour market risks. The present system does not provide adequate incentives to prevent or mitigate potential unemployment through voluntary mobility. However, new securities are a precondition for encouraging riskier decisions such as the move from waged work to self-employment or even to embark on the venture of changing occupation once or maybe twice in the life-course. The old insurance systems are still loaded with elements of the traditional gender division. Internal labour markets as an insurance device are eroding. Functionally equivalent insurance systems against earnings risks over the life-course, however, are not yet in place or are only just beginning to emerge.

For this reason, some form of employability or work-life insurance should supplement the time-honoured system of unemployment insurance. This should cover income risks not only in the worst case – that is, unemployment – but also during risky transitions into new employment relationships.

Three prominent examples will be given that might be considered good practice for work-life insurance. The first relates to the transformation of job protection into a kind of mobility insurance.

1. Two years ago, Austria changed its severance pay law (*Abfertigungsrecht*) into a kind of income protection in the event of job mobility. The law was initiated by the social partners (!) and aimed at responding to increasing job instability. In the event of dismissal, each employee receives severance pay or can leave it on account and claim it later, even

if she has only a brief employment record or quits voluntarily. The amount of severance pay increases with the number of years employed: e.g. two months' salary for three years' employment to 12 months' salary for 25 years' employment. Financing comes from employers' contributions of 1.53 per cent of the salary. There is no severance pay for less than three years' employment, or for wrong-doing which results in resignation or dismissal. However, in these cases, the entitlement remains on the account, until it becomes a legitimate claim. A special 'Employee Accounting Office' (*Mitarbeiterversorgungs-Kasse*) administers the fund. The insurance claim, therefore, is directed to the *Versorgungskasse* and not to the individual employer. Employees have the choice of cash payment or of an additional contribution to their pension fund.

The Netherlands has also sector-specific severance funds (*Wachtgeld*) financed by employers and incorporated in its unemployment insurance system (from 0.07 per cent contributions of wages in the insurance economy to 2.78 per cent in cultural services).

The Danish system is a functional equivalent to such mobility insurance. It combines a low level of job protection with a high level of transfer payments in the event of unemployment at least for low and medium income earners. About 40 per cent of unemployed workers eventually return to the same employer.

2. The second example is the concept of wage insurance. There are two programmes, one (good practice case) in Switzerland and one (less good practice case) in Germany. The concept was discussed during the Clinton administration in the US but has not been adopted by the Bush administration. However, the reasons for introducing such a scheme in the US are interesting and can be generalized:

- There is a real need for mature adults who face income losses of up to 30 per cent in the event of unemployment and subsequent re-employment.
- The scheme subsidizes training on the job which is much more effective than training off the job where re-employment prospects are uncertain.
- Wage insurance payments could be complemented with training vouchers to make new jobs sustainable or to improve the likelihood of advancing on the career ladder.
- Wage insurance reduces workers' opposition to trade liberalization, which is, according to surveys, high where there is no assistance; estimated costs of about $4 billion would be a tiny fraction of the $500 billion in estimated benefits for the US from freer trade.

The Swiss scheme of *Zwischengeld* (compensatory benefit) works as follows. If a job is accepted that pays less than a person's unemployment compensation (ALE), the unemployment insurance fund pays a compensatory benefit so that this and the earnings from that 'temporary' job together exceed the unemployment compensation. For example:

Insured wage:	SFr. 5,000
ALE:	SFr. 4,000 (80 per cent of insured wage)
Zwischenverdienst:	SFr. 3,000 (less than ALE)
Compensation	SFr. 1,600 ((insured wage − *Zwischenverdienst*) × benefit rate 80 per cent)

The entitlement to the *Zwischenverdienst* is limited to the first 12 months. For insured persons with dependent children and insured persons aged over 45 years the maximum period is two years. The period during which *Zwischenverdienst* is earned constitutes new contribution periods counting towards the qualifying period. An evaluation study on the basis of rigorous methods and suitable data has given this scheme the best scores in terms of efficiency and effectiveness.

Following the proposal of the Hartz Commission, the German government introduced wage insurance for unemployed people over the age of 50. However, the programme was so badly designed that the take-up is very low.[17]

3. The third example is Dutch life-course saving plan. According to this, workers can save 12 per cent or their annual income tax-free up 210 per cent of their annual salary. Workers can use the saving account to finance unpaid leave during their working life, or early retirement. During leave-of-absence, they are eligible for an additional tax cut. This means that after saving 12 per cent over 17.5 years, the maximum saving account is reached (17.5 x 12 = 210 per cent). In this way, workers can finance three to four years' unpaid leave at 70 per cent of their usual wage. Social partners can make collective arrangements with banks or insurance companies in the Collective Labour Agreements. These agreements are not mandatory.

The Dutch life-course plan, however, is not without its critics and has some flaws from the TLM point of view, despite its bold and – due to its simplicity – persuasive character. In the present design of income-contingent tax incentives, high-paid workers are favoured, and there is no provision for leave for continuous training or for upgrading the low-skilled. It is to be expected that the present scheme will be taken up mainly for compensating income loss at early retirement.

Conclusion

European integration created a fundamental asymmetry between policies promoting market efficiency at the European level and those promoting social protection and equality at the national level. The debate on the European social model, therefore, is a debate on Europeanizing social policy (Scharpf 2002). In reality and in the medium term, social policy in Europe will be shaped according to the various welfare traditions and to the different stages of economic development. However, the open method of coordination – understood not only as a cognitive but also as a normative process – might help to shape something like a 'European social model' in the long run (Rodrigues 2002: 22–3). In this way, Europe might even move to a confederation of mutually learning and supporting societies.

The contribution to this debate here is the proposal for an encompassing work-life insurance based on the concept of transitional labour markets. It could serve as a regulatory idea for social justice over the life-course as well as for a work-life balance to endorse overall gender equality in the labour market and to increase freedom of choice. The social-liberal vision of 'asset equality' or of the 'social investment state' cannot substitute for the need for continuous redistribution during the life-course in favour of people affected by old and especially new risks. Enabling workers alone does not suffice since bad luck can put anyone in a desperate situation despite their talent and individual effort. Thus, the modern welfare state has to ensure solidarity – out of compassion and to avoid inequality – through periodic redistribution of resources and through re-establishing individuals' capabilities to manage their life. The concept of social risk management through transitional labour markets provides a modern and realistic strategy for equality and an efficiency-enhancing strategy, emphasizing both individual and collective responsibility for social risks. Needless to say, it requires further rigorous empirical, analytical and normative foundations.

Notes

1. See *inter alia* Ferrera et al. (2000), Zeitlin and Trubek (2003) and Kaelble and Schmid (2004).
2. See, for instance, Giddens (1998).
3. See, for instance, Considine (2001), Hoffmann-Riem (2001) and Schuppert (2003).
4. For a more extensive treatment of the development of new employment risks, see Esping-Andersen (1999) and Schmid (2005).

5 These and the following figures are taken from Kok et al. (2004), European Commission (2002–2004) and OECD (2004).
6 For the first fundamental study on fixed-termed contracts in a comparative perspective, see Schömann et al. (1998).
7 See de Bruijn et al. (2003) and Macdonald and Holm (2001).
8 Sweden's high employment rate (69 per cent) of mature workers is explained by at least four institutional arrangements fitting perfectly within the TLM framework: 1) intensive training for mature workers (among others through the massive *'knowledge lift programme'* of 1997–2002); 2) the dismantling of any monetary incentives of early retirement; 3) the establishment of 'work-adjustment groups' in firms with more than 50 employees who are responsible for relocating or rehabilitating mature workers if they experience reduced income or work capacities. 4) gender-related differences in mandatory retirement (and probably the institution of mandatory retirement at all) are out of date. Women should accumulate pension rights independent of the working career of their ('breadwinning') spouse. This is a feature that Sweden (as well as Switzerland) has established.
9 See also Holzmann and Jorgensen (2000) who make, among other points, a useful distinction between preventing, mitigating and coping with risks.
10 For normative considerations of social risk management (risk-sharing) and the methodology of risk assessment, see Schmid (2005).
11 For good practice of risk analysis in the vein of transitional labour markets, see O'Reilly et al. (2000), de Koning and Mosely (2002), Schömann and O'Connell (2002), Muffels et al. (2002), Meager and Bates (2002) and Gangl (2003); for a special emphasis on the gender dimension cf. Mosley et al. (2002)
12 For the productivity-enhancing effects of unemployment (and respectively income inequality) insurance probably exceeding the related moral hazard effects, see Acemoglu and Shimer (2000) and Gangl (2003).
13 It can be debated whether a basic income guarantee would be necessary to complement the system of 'employment insurance' as a fourth pillar (see Standing 1999). It makes sense, in my view, to complement pension insurance with such a guarantee, as is already the case in some countries (e.g. the Netherlands, Denmark and Sweden). The TLM framework, however, provides work-related income guarantees at several stages of critical transitions through in-work benefits or accumulated flexible drawing rights (e.g. flexible pension entitlements).
14 Individual preferences could be enhanced even more strongly by allowing vouchers or drawing rights to be exchanged, within companies or perhaps in some sort of 'voucher exchange' (Wilthagen and Rogowski 2002).
15 This slogan has been coined by Gazier (2002, 2003).
16 Quoted in Bernstein (1996), p. 119f.
17 After dismissal from work, the German wage insurance (as a supplement to the unemployment insurance system) replaces part of the loss of income suffered by older people for the first few years if they take less well paid jobs subject to compulsory social insurance. The benefit may be accumulated with wage cost subsidies for the employer. The scheme, however, is not well designed according to the rules of TLM. The wage compensation will be paid only during the remaining time of unemployment benefit entitlements. If the unemployed person has exhausted their benefit entitlement, they will not receive any more a compensation; if they have exhausted 50+ per cent of the benefit period

(say, 12 months of a maximum period of 18 months), they will receive compensation only for the rest of the UB entitlement (i.e. in the example for six months). Since most older people – especially in East Germany – are long-term unemployed, the wage insurance incentive has had a very limited impact.

References

Acemoglu, Daron and Robert Shimer (2000) Productivity Gains from Unemployment Insurance. *European Economic Review*, 44: 1195–1224.
Bernstein, Peter L. (1996) *Against the Gods: The Remarkable Story of Risk* (New York: John Wileys & Sons).
de Bruijn, Jeanne, de Vries, Gert and ten Cate, Helen (2003) The Burden of Late Modernity: Work-Related Disability among Women in the Netherlands, unpublished manuscript.
Considine, Mark (2001) *Enterprising State:. The Public Management of Welfare-to-Work* (Cambridge: Cambridge University Press).
Considine, Mark, Brian Howe and Lauren Rosewarne (eds.) (2005) *Transitional Labour Markets in Australia* (Oxford: Oxford University Press).
Dworkin, Ronald (2000) *Sovereign Virtue. The Theory and Practice of Equality* (Cambridge, MA and London: Harvard University Press).
Esping-Andersen, Gösta (1999) *Social Foundations of Postindustrial Economies* (Oxford: Oxford University Press).
European Commission (2002–2004) *Employment in Europe 2002, 2003, 2004 – Recent Trends and Prospects* (Luxembourg: Office for Official Publications of the European Communities).
Ferrera, Maurizio, Anton Hemerijck and Martin Rhodes (2000) *The Future of Social Europe: Recasting Work and Welfare in the New Economy*. Oeiras: Celta Editora.
Gangl, Markus (2003) *Unemployment Dynamics in the United States and West Germany. Economic Restructuring, Institutions and Labour Market Processes* (Heidelberg and New York: Physica/Springer).
Gazier, Bernard (2002) Transitional Labour Markets: From Positive Analysis to Policy Proposals, in Günther Schmid and Bernard Gazier (eds.) *The Dynamics of Full Employment: Social Integration through Transitional Labour Markets* (Cheltenham and Northampton, MA: Edward Elgar), pp. 196–232.
Gazier, Bernard (2003) *'Tous Sublimes'. Vers un nouveau plein-emploi* (Paris: Flammarion).
Giddens, Anthony (1998) *The Third Way: The Renewal of Social Democracy* (Cambridge: Polity Press).
Hoffmann-Riem, Wolfgang (2001) *Modernisierung von Recht und Justiz. Eine Herausforderung des Gewährleistungsstaates* (Frankfurt a.M.: Suhrkamp).
Holzmann, Robert and Jorgensen, Steen (2000) *Social Risk Management: A New Conceptual Framework for Social Protection and Beyond* (Washington: Social Protection Discussion Paper Series No. 0006 (The World Bank)).
Howe, Brian, Renate Howe, Andrew Burbidge, Jan Manning, Julia Perry and Peter Stricker (2005) *Transitions in Australian Labour Markets: Initial Perspectives* (Melbourne: CEDA), Information Paper 82.
Kaelbe, Hartmut and Günter Schmid (eds.) (2004) Das europäische Sozialmodell: Auf dem Weg zum transnationalen Sozialstaat, *WZB-Jahrbuch 2004*. Berlin: edition sigma.

Kahneman, Daniel and Amos Tversky (eds.) (2000) *Choices, Values and Frames* (Cambridge: Cambridge University Press).
Kletzer, Lori G. and Robert E. Litan (2001) *A Prescription to Relieve Worker Anxiety*. Policy Brief 01-2, Institute for International Economics and Brookings Institution, February 2001; http://www.iie.com/publications/pb/pb01-2.htm.
Kok, Wim et al. (2004) *Jobs, Jobs, Jobs: Creating More Employment in Europe*, Report of the Employment Task Force Chaired by Wim Kok, Luxembourg: Office for Official Publications of the European Communities. www.europa.eu.int/comm/employment_social/ employment_strategy/task_en.htm
de Koning, Jaap and Hugh Mosley (eds.) (2001) *Labour Market Policy and Unemployment: Impact and Process Evaluations in Selected European Countries* (Cheltenham and Northampton, MA: Edward Elgar).
Lechner, Michael, Markus Frölich and Heidi Steiger (2004) *Mikroökonometrische Evaluation aktiver Arbeitsmarktpolitik*. Abschlussbericht für NFP 43, Projekt-Nr. 4043-058311 (Sankt Gallen: Universität St Gallen).
Macdonald, Fiona and Sonya Holm (2001) Employment for 25- to 34-Year-Olds in the Flexible Labour Market: A Generation Excluded? in Linda Hancock, Brian Howe, Marion Frere and Anthony O'Donnell (eds.) *Future Directions in Australian Social Policy. New Ways of Preventing Risk* (Melbourne: CEDA).
Meager, Nigel and Peter Bates (2002) From Salary Workers to Entrepreneurial Workers? in Günther Schmid and Bernard Gazier (eds.) *The Dynamics of Full Employment: Social Integration through Transitional Labour Markets* (Cheltenham and Northampton, MA: Edward Elgar), pp. 298–339.
Mosley, Hugh, Jacqueline O'Reilly and Klaus Schömann (eds.) (2002) *Labour Markets, Gender and Institutional Change: Essays in Honour of Günther Schmid* (Cheltenham and Northampton, MA: Edward Elgar).
Muffels, Ruud, Ton Wilthagen and Nick van den Heuvel (2002) *Labour Market Transitions and Employment Regimes: Evidence on the Flexicurity–Security Nexus in Transitional Labour Markets*. WZB Discussion Paper FS I 02-204 (Berlin: Wissenschaftszentrum Berlin für Sozialforschung).
Newgart, Michael and Klaus Schömann (eds.) (2002) *Forecasting Labour Markets in OECD Countries* (Cheltenham and Northampton, MA: Edward Elgar).
OECD (2004) *Employment Outlook* (Paris: OECD Publications).
O'Reilly, Jacqueline, Immaculada Cebriá and Michel Lallement (eds.) (2000) *Working-Time Changes. Social Integration through Transitional Labour Markets* (Cheltenham and Northampton, MA: Edward Elgar).
Rawls, John (1990) *A Theory of Justice* (Oxford: Oxford University Press).
Rawls, John (2001) *Justice as Fairness – A Restatement*, ed. Erin Kelly (Cambridge, MA and London: The Belknap Press of Harvard University Press).
Rodrigues, Maria J. (ed.) (2002) *The New Knowledge Economy in Europe* (Cheltenham and Northampton, MA: Edward Elgar).
Scharpf, Fritz W. (2002) The European Social Model: Coping with the Challenges of Diversity. *Journal of Common Market Studies*, 40: 645–70.
Schmid, Günther (2002) Transitional Labour Markets and the European Social Model: Towards a New Employment Pact, in Günther Schmid and Bernard Gazier (eds.) *The Dynamics of Full Employment. Social Integration through Transitional Labour Markets* (Cheltenham and Northampton, MA: Edward Elgar), pp. 393–435.

Schmid, Günther (2005) Social Risk Management through Transitional Labour Markets: Theory and Practice Related to European Experiences, in Marc Considine, Brian Howe and Lauren Rosewarne (eds.) *Transitional Labour Markets in Australia* (Oxford: Oxford University Press).

Schmid, Günther and Bernard Gazier (eds.) (2002) *The Dynamics of Full Employment. Social Integration through Transitional Labour Markets* (Cheltenham and Northampton, MA: Edward Elgar).

Schömann, Klaus, Ralf Rogowski and Thomas Kruppe (1998) *Labour Market Efficiency in the European Union: Employment Protection and Fixed-Term Contracts* (London and New York: Routledge).

Schömann, Klaus and Philip J. O'Connell (eds.) (2002) *Education, Training and Employment Dynamics. Transitional Labour Markets in the European Union* (Cheltenham and Northampton, MA: Edward Elgar).

Schuppert, Gunnar F. (2003) *Staatswissenschaften* (Baden-Baden: Nomos).

Sen, Amartya (2001) *Development as Freedom* (New York: Alfred A. Knopf).

Shiller, Robert (2003) *The New Financial Order: Risk in the 21st Century* (Princeton, NJ: Princeton University Press).

Standing, Guy (1999) *Global Labour Flexibility: Seeking Distributive Justice* (London and New York: Macmillan and St. Martin's Press).

Supiot, Alain (2001) *Beyond Employment. Changes in Work and the Future of Labour Law in Europe* (Oxford: Oxford University Press).

Wilthagen, Ton and Ralf Rogowski (2002) Legal Regulation of Transitional Labour Markets: Barriers and Opportunities, in Günther Schmid and Bernard Gazier (eds.) *The Dynamics of Full Employment. Social Integration through Transitional Labour Markets* (Cheltenham and Northampton, MA: Edward Elgar), pp. 233–273.

Zeitlin, Jonathan and David M. Trubek (eds.) (2003) *Governing Work and Welfare in a New Economy: European and American Experiments* (Oxford: Oxford University Press).

Part II
Security Policies: Assessing and Mitigating Risks

Part II

6
Security Policies in a Transnational Setting
Ursula Schröder

Introduction

After the demise of the Cold War, security provision stopped being synonymous with the military defence of the nation-state. Today, the security agenda is populated by a broad selection of transnational security risks, ranging from the trafficking of drugs and migrants to terrorism and the spread of infectious diseases. Faced with this complexity, individual states are less and less able to offer effective solutions to boundary-transcending challenges and those actors responsible have to design new political strategies that deal with the radical transformation of today's security environment. In search of a new paradigm, political and academic security debates frequently resort to the concept of 'risk management' to replace traditional security terminology coined during the Cold War. Yet in many cases it is used without any specification of its substantive content or the larger implications for the politics of security. In order to fill this void, this chapter gives an overview of the, albeit as yet limited, state of the art on risk debates in security research and discusses their analytical and normative implications. It examines what repercussions the introduction of risk terminology has had on the development of security political strategies and where the limits of transferring strategies of risk management to the security sector lie. The following discussion of risk political strategies that respond to the rise of transnational security challenges proceeds in three steps. First, the concept of 'risk' is adapted to the field of international security and furnished with a range of examples explaining its use. Second, possible strategies of political risk management are discussed, exploring the territory of proactive risk management strategies and their corresponding comprehensive security solutions. The concluding case study examines the politics and organization of European crisis management in relation to the debated reconfiguration of the European security field.

International security: from threats to risks

The concept of risk currently enjoys great popularity in the field of international security and has become a catch-all phrase for the description of an increasingly complex post-Cold War security environment. Debates in academic and policy circles about what constitutes a security challenge pushed the boundaries of a traditionally state-centric and military-based security agenda. But although the concept of risk today is widely used as a replacement for the threat/deterrence model of international security that characterized the Cold War era, it remains underdetermined in its substantive content. Risk, in everyday parlance, tends to be negatively connoted as 'hazard' with an emphasis on adverse 'chance' events or accidents, and is thereby defined as the opposite of 'opportunity'. In contrast to this notion of 'unknowable hazard', the professional debate defines risk in terms of probability: risk is characterized as a function of the relative likelihood of a future event coupled with its potential damage. In this case, stakeholders know the statistical likelihood and nature of a harmful event taking place in the future and decide on the basis of this mathematical risk assessment on the course of action to take. In risk sociology, this definition is expanded: sociological risk analyses assume that human agency has an impact on both the likelihood and amount of harm induced by a specific risk (Luhmann 1991: 30f). Risk is a product of our own action and is therefore reflexive (Rasmussen 2004: 393). The larger implications of the reflexive concept of risk became apparent in Ulrich Beck's famous phrase 'risk society' being a (if not the) distinctive characteristic of the second or 'reflexive' period of modernity (Beck 1992; 1999). Risk society refers to 'a phase of development of modern society in which the social, political, ecological and individual risks created by the momentum of innovation increasingly elude the control and protective institutions of industrial society' (Beck 1999: 72). From being a technical term denoting the calculable probability of future harm, risk became the condition of modern society that itself creates the hazards and insecurities it has to deal with. In this perspective, our own actions influence the occurrence and scope of future risks. Moreover, the calculation of risks 'is contingent on rules and norms, institutions and interests. Risks are not "simply out there" to be measured' (Daase 2002a: 12). The shared societal standards about what constitutes a risk and where to fix thresholds for acceptable levels of risk vary across societies and throughout history. Potentially dangerous events and developments continually surface and disappear in public awareness, whilst attempts at political regulation are made to respond to these tides. Thus, a risk is what we define as a risk. This holds true for the security sector, in which the traditional 'threat paradigm' has given way to a widened focus on a whole set of risks that can affect the security of states, societies or individuals (see Wæve et al. 1993; Suhrke 1999). The traditional configuration of security threats serves as a yardstick against which to delineate the novelty of international security risks. In the classical 'narrow' definition of national security, as a situation in which military threats to the state are absent (Wolfers 1962), a security threat is characterized by three basic conditions: 1) the existence of an adversarial actor – be it a state or a sub-state collective – combined with 2) hostile intentions and 3) the likelihood that this actor has the military potential to inflict noteworthy harm (Daase 2002a: 15). This 'threat triangle' served as the main organizing principle of security provision during the Cold War. Meanwhile, however, it lost a lot of its explanatory power due to the radical transformation of the security environment. In one of the few research programmes dealing conceptually with risks in international relations, Daase et al. (2002a) argue that traditional threat paradigms have become obsolete in the face of the uncertainty pervasive in international security politics. Whereas threats were linked to hostile intentions, military capabilities and adverse actors, security risks are defined as lacking at least one of these conditions (Daase 2002a: 16). The certainty of preparing for defined, intentional and obvious security threats gives way to a multitude of potential risk factors, often devoid of clear assignments of specific actors or intentions.

Most contemporary security risks are both transnational and transversal in scope. Transnationality refers to the phenomenon that most security risks move across national borders. Security challenges like terrorism, organized crime or environmental disasters often affect several states simultaneously (Politi 1997; Treverton 2005). Similarly, the rise of transnationally organized non-state actors in the security sector constitutes a fundamental departure from an international system that accepted only sovereign states as security actors (see Adamson 2002; Williams 2002). In addition to their transnational character, new security risks increasingly traverse the boundaries between the internal and external security sectors, a fact that has only recently started to gain attention in security debates (see Andreas and Price 2001; Bigo 2001). The case of terrorism is a salient example of this dual trend. Some forms of terrorist activities have moved beyond territorial limits, and terrorists are now organizing in transnational networks of semi-autonomous cells. The fight against this rapidly globalizing phenomenon often involves both military and law enforcement organizations and merges the tasks of the domestic and foreign intelligence services, thereby cutting across traditional organizational entities (see Beck 2002; Coker 2002: 39f; Daase 2002b; Schneckener 2002).

The described make-up of the new security agenda brings with it uncertainty about the nature of possible harm and about viable strategies of risk management. Security strategies can no longer offer finite solutions to clearly specified security challenges; instead they are left with managing insecurity, not with providing security (Coker 2002: 62). The management of an increasingly complex and opaque risk environment engenders a double transformation of security provision: security policies move towards the proactive curtailment of future harm and devise comprehensive strategies with which to confront transnational and cross-cutting security challenges.

Towards the proactive management of risk

Proactive strategies of risk management attempt to limit the uncertainty of the new security environment by promoting a range of preventive and precautionary policies that deal with security risks before they escalate. This constitutes a fundamental departure from the traditional understanding of security provision as a reactive 'defence of the state', based on the relative stability of balance-of-power conceptions of order in the international system of states. The armed and police services, formerly occupied with external 'territorial defence' and the domestic 'aversion of imminent danger', currently find themselves in situations in which they are requested to act before a specific danger to their society has materialized. The rationale behind this is that due to incomplete information about the scope and sometimes even the nature of future harm, security policies have become an exercise in preparedness against a broad range of risk scenarios with particular emphasis on reducing or preventing possible harmful outcomes to society.

The two main forms into which proactive security policies can be categorized are preventive and precautionary strategies. Although the terms are often used interchangeably, their substantive content is quite different. Preventive security policies seek to avert catastrophes by thwarting the likelihood of future harm. The 'precautionary principle', borrowed from the environmentalist discourse of the 1990s (see Houben 2003: 2f), does not share the negative connotations of the 'prevention paradigm'. It is essentially not interested in preventing a catastrophe from happening, but is instead based on the positive notion of preserving a particular common good. Precautionary politics frequently deal with the environment or the general health of the public.[1] The linchpin of the argument is to call for precautionary action in cases where, even if the proof of future damage is inconclusive, waiting for higher levels of proof could be too costly. The key difference between precautionary and preventive strategies lies in their diverging understanding of acceptable levels and burdens of proof: preventive action relies on the certainty of imminent danger and the proof of 'sufficient threat' (Houben 2003: 11). Hence, 'in case of pre-emptive military action, the burden of proof is with the actor contemplating the strike' (ibid.). In precautionary action, the level of proof is reversed: the evidence-based legitimation of actions becomes unnecessary, since the principle of precaution applies precisely in situations of outstanding proof or uncertainty about the causal links between action and damage.

Applying these principles to the security sector is not straightforward, since the traditionally defensive organization of the security sector does not lend itself easily to preventive or precautionary action. Moreover, proactive security policies often take place beyond the normative and legal limits of what is domestically or internationally deemed to be an acceptable security strategy. With organizational problems meeting normative limitations, prevention and precaution provide no easy solutions to the problems of a complex security agenda. The following range of examples outlines the challenges of the concept.

The strategy of preventive intervention is a prototypical example of the normative and legal problems that result from proactive security policies. Preventive or 'pre-emptive' interventions are based on the principle of attacking an inimical state first, if there is reliable proof that aggression by this adversary is imminent. Those criticizing preventive interventions object to their inherent violation of international law, and in particular to the principle of non-intervention into the internal affairs of sovereign states and the convention against waging a war of aggression (see Bothe 2003). Proponents of preventive interventions put forward arguments about their necessity in light of a changed security environment in which reactive security strategies can lead to catastrophic consequences on the scale of massive terrorist attacks or worse (see Sofaer 2003). The classic textbook case of a preventive external security strategy is the Israeli defence doctrine that led to a pre-emptive strike against Egypt at the beginning of the 1967 Six-Day War (see Creveld 1998; Schiff 1999). Other examples include the recent intervention into Iraq as well as cases of military counter-proliferation against states supposedly in the possession of weapons of mass destruction. Here, the most commonly cited example is the contentious Israeli strike against Iraq's Osirak nuclear reactor in 1981. This preventive strike was later declared illegal by the UN Security Council, which claimed that the Israelis had threatened regional security.

The debate about applying the precautionary paradigm to the security sector is inconclusive and there is little overall agreement on what it could mean for the provision of security. One of the few consensual justifications for precautionary action applies to cases in which security stakeholders prepare for domestic emergencies. Civil protection and consequence management programmes are designed to minimize the impact of natural and man-made disasters such as floods, earthquakes, the spread of infectious diseases or the aftermath of terrorist attacks. The underlying logic is that due to uncertainty about future emergencies, precautionary measures have to be taken to reduce both their impact and their likelihood. In recent years, precautionary strategies related to the protection of the 'homeland' against terrorist attacks have had priority in both Europe and the US.[2]

More controversial are precautionary security measures that are directly related to preventing specific criminal activities. Particularly decision-makers in the field of counter-terrorism argue that the reactive and prosecution-based character of traditional policing is insufficient. The shift from a prosecution-based to a preventive model of policing implies in this case that the law enforcement agencies can preventively investigate terrorist crimes that have not yet been committed in order to avert danger from society. This proactive management of risks of terror attacks can take the form of 'policing communities of risk' (Johnston 1997; see also Ericson and Haggerty 1997). Precautionary measures, particularly the expansion of internal surveillance, are in this case taken against specific target groups that have a statistically higher likelihood of perpetrating certain crimes. Critics accuse this form of risk management of constituting an unacceptable infringement on civil liberties in which the aversion of dangers to society leads directly to a 'surveillance' or 'police' state (see Walter et al. 2004 for the legal debate). A recent and widely debated example is the UK Prevention of Terrorism Act 2005: the moot point was the introduction of 'control orders' curbing the activities of suspected terrorists, which range from tagging suspects to placing them under what is effectively house arrest. The trade-off between security and civil liberties is particularly evident in this case, where police intervention is legitimized only by intelligence sources, not by evidence admissible in court, and where crimes have not yet been committed.

Coping with transnational and cross-cutting risks

Complementary to the proactive management of security risks, the design of comprehensive security strategies provides answers to the rise of transnational and transversal security risks that obliterate the customary national and sectoral divisions of security provision. Current security strategies, despite ongoing misgivings about a perceived lack of analytical clarity and strategic poignancy in multifaceted concepts of security (cf. Walt 1991; Baldwin 1997; Buzan, Wæver and de Wilde 1998 for the debate) cover a broad range of civilian and military instruments to generate comprehensive forms of security provision. A good example is the European Security Strategy 'A Secure Europe in a Better World', presented by Javier Solana, High Representative for the EU's Common Foreign and Security Policy and adopted by the Heads of State at the European Council on 12 December 2003. However, new conceptual and organizational solutions to the challenges of transnationality and transversality do not come easily and in both cases the problems remain largely unresolved.

Whereas social and political life used to be based on nation-states and generally was understood in a territorial sense (Beck, 1999: 1f), the new global risks have rendered the exclusively national provision of security ineffective. Accordingly, political answers more and more often seek solutions beyond the territorial organization of security provision. In Beck's terms, the *'national organization as a structuring principle of societal and political action can no longer serve as a premise for the social science observer perspective'* (Beck 2002: 52, emphasis in original). Against the necessity to cooperate internationally stand the primacy of national sovereignty in matters of security and its institutional representation: the state monopoly on violence (Tilly 1975; Giddens 1985). This principle of national primacy derives from the realist assumption of a security dilemma inherent in the anarchical international system of states that impedes attempts to tackle the new security challenges in a cooperative fashion (Wheeler and Booth 1992). In the case of the European Union, the problems of the historical limitation of security policies to the realms of nation-states are obvious and the gap between political rhetoric and the actual implementation of cooperative measures is particularly pronounced: security cooperation is to this day hampered by a lack of willingness on the part of EU member states to cede their exclusive sovereignty over security issues.

Simultaneously, the emergence of cross-cutting horizontal security issues leads to the transformation of the traditional bifurcated state monopoly on violence. Whereas previous security challenges more or less fitted into the categories of internal and external security, and thus fell into the respective fields of competence of the different internal and external security services, this distinction can no longer be made without neglecting important aspects of the new risks. In the case of the European Union, the organizational problems caused by cross-cutting security issues are particularly pronounced: current institutional venues are not designed to deal with transversal security challenges, since the existing European security architecture mirrors the traditional separation of security affairs into internal and external aspects. Hence, the provision of security is partitioned among different agencies and departments that often compete with each other for scarce competences, resources and issue definitions. Innovative solutions to complex security problems that involve several agencies and departments are hindered by these in-built conflicts.

Yet, despite the difficulties accompanying the organizational forms and institutional cultures of the security agencies, the reconstruction of traditional security provision towards vertical international security cooperation and horizontal inter-departmental security coordination is under way. This twofold phenomenon constitutes a turning-point in the organization of EU security provision after the end of the Cold War. The following study of EU crisis management policies highlights how the challenge of both vertical and horizontal integration of security political strategies is tackled in the European Union.

EU crisis prevention and management

Recent EU policies on conflict prevention and management constitute a good case for exemplifying the challenges of security policies in a transnational setting, since they combine complex environmental conditions with a proactive and comprehensive security strategy. In line with a broad understanding of security, the European Security Strategy argued that 'Europe faces new threats which are more diverse, less visible and less predictable' (p. 3) than former security challenges and featured the security risks of state failure and regional conflicts as two of the five most relevant challenges currently facing Europe.[3] Accordingly, 'bad' or 'weak' governance structures[4] are seen as the underlying problem in many 'countries at risk of instability', as a recent high-level study in Great Britain cautiously phrased the problem (Strategy Unit, 2005).

They can result in both civil and violent conflicts that corrode states from within (see Migdal 1988; Zartman 1995; Milliken 2003). Yet, governance problems have severe repercussions not only for the directly affected societies, but also for neighbouring states and regions. They often leave terrorism and organized crime networks to grow unchecked and moreover increase the likelihood of violent conflicts and instability spilling over into other states. The EU sees the negative impact of external regional instabilities and societies at risk as one of its main security challenges and argues that 'neighbours who are engaged in violent conflict, weak states where organized crime flourishes, dysfunctional societies or exploding population growth on its borders all pose problems for Europe' (European Security Strategy: 7f).

European political strategies dealing with the problems of regional instability and conflict are mostly of a proactive nature, as implied by the term 'conflict prevention'. Besides a general moral obligation to act in response to civil strife or large-scale human rights abuses (see Walzer 1977; Hoffmann 1995; Wheeler 2000), their main aim has been broadly defined as the 'stabilisation and pacification of all those states in which Europe perceives itself to have moral and practical concerns' (Hill 2001: 316). Strategies of conflict prevention and management can thus be understood as being 'fundamentally about the effective management of risks of instability to reduce the occurrence of crisis' (Strategy Unit 2005: 6). Fostering 'structural stability'[5] became a primary goal in order to reduce the negative impact of 'countries at risk', particularly in the EU's near-abroad. Understood first and foremost as the promotion of good governance, the European search for stability is summed up in the idea enshrined in the European Security Strategy that 'the best protection for our security is a world of well-governed democratic states' (p. 10).

The stability-oriented strategies pursued by the EU vary according to their time horizon and their proactive or reactive character. Proactive strategies are generally divided into two classes of short-term 'operational' and longer-term 'structural' types of conflict prevention (Ackermann 2003: 341). The EU external policies of development cooperation, external assistance, economic cooperation and human rights policies are long-term structural policies geared at eliminating the root causes of violent conflict. Short-term policies, on the other hand, are often eleventh-hour attempts at de-escalating already virulent conflicts through preventive diplomacy (Lund 1996) or preventive military deployment (Ackermann and Pala 1996; Ackermann 2000). The EU first codified its goal of developing conflict prevention capacities in the 'Programme on the Prevention of Violent Conflict' adopted at the European Council in Gothenburg 2001 and later in the European Security Strategy, which named conflict prevention as a strategic priority.

In complex security environments, however, prevention is no universal remedy and recent progress in the European Union's civilian and military crisis management capabilities bears witness to the increased pertinence of 'second-generation' or 'post-conflict' prevention and peace-building (Schnabel 2002). The rapid institutionalization of a European crisis response capability following the Cologne European Council in June 1999 (Howorth 2001; Deighton 2002) enables the EU to undertake a limited number of autonomous civilian and military peace support operations. So far, military missions have been deployed to Macedonia (Concordia), Bosnia-Herzegovina (Althea) and the Democratic Republic of Congo (Artemis). Civilian and non-executive police missions are or were active in Macedonia (Proxima), Georgia (EUJUST Themis), Bosnia-Herzegovina (EUPM) and the Democratic Republic of Congo (EUPOL Kinshasa) (see Cussen 2004; International Crisis Group 2005). In sum, the EU's goal of fostering regional stability and preventing the impact of instability and conflict on European territory is served by a whole range of different policy instruments. Progress in the institutionalization of the 'preventive paradigm' in European security policy has been reasonably fast in recent years. But a major remaining challenge is the coordination of both short- and longer-term initiatives and the overall coherence of integrated European crisis management and peacebuilding missions.

Due to the range of instruments at its disposal, the European Union is widely seen to be in a unique position to provide an integrated approach to conflict management. Yet, while policymakers have stressed the singular potential and added value of the EU in the fast-growing area of peace support missions (see European Security Strategy: 14 and COM (2001) 211: 9), the actual pursuit of policy coherence is still beset with a range of problems. The traditional pillar structure of EU policymaking is particularly counter-productive to the formulation of coherent policy initiatives, since diverse and sometimes conflicting political structures and organizational cultures slow down the pace of progress. The historical emergence of the EU treaty base resulted in the division of security political competences to first pillar 'community activities' in the fields of development cooperation, human rights, humanitarian and trade policies, second pillar Common Foreign and Security Policy (CFSP) or third pillar Justice and Home Affairs (JHA) actions. This delegation of security policy issues into different internal and external decision-making arenas comes at the expense of comprehensive security political answers.

Assuming that organizational form influences policy outcomes, 'problems falling between jurisdictions, or cutting across jurisdictional boundaries, are at a disadvantage in the competition for policy attention' (Scharpf 1986: 181f). This is the case for complex security issues that sit uneasily between the traditional arenas of political decision-making, since they concern several security agencies at the same time. Coordination problems arise particularly in areas where the distribution of competences between the different departments is contested. Recent years have witnessed several such interfaces and 'occupational overlap' problems between the actors active in European peace support initiatives. Two examples will serve to indicate some of the larger debates and problems surrounding the emergence of comprehensive security strategies: the relationship between short-term and long-term conflict prevention and the relationship between different actors active in short-term peace support operations.

In the EU, structural and long-term development assistance projects are first pillar activities. They are enshrined in the Commission's country or regional strategy papers that highlight the areas in which community projects can make a significant contribution to the stability and recovery of the country or region in question.[6] Short-term crisis management capabilities, on the other hand, were developed within the European Council structures in the context of second pillar European Security and Defence Policy (ESDP) mechanisms.[7] In contrast to the longer-term and project-based approach of developmental policies, interventionist crisis management takes the form of high-profile and often politically symbolic civilian and military operations. The ongoing challenge is to link crisis management with conflict prevention in a way that enables the different organizations and agencies to make a concerted effort in dealing with the interlinked socio-economic, political and security dimensions of the conflict in question. But as a result of the unhelpful pillar structure that instils competition about tasks and resources between the various agencies involved, inter-institutional relations are generally weak or, at best, rely on informal interactions due to a lack of formalized inter-institutional venues. The relationship between the Council and the Commission is an exemplary case of the failure of integrating short-term operations into longer-term country strategies, since the Commission deals exclusively with longer-term community projects, while the Council manages short-term operations. Due to this institutional separation of crisis management operations from the wider range of conflict prevention strategies, Council–Commission relations are difficult by design and often strained. The problematic division of competences between the intergovernmental Council structures and the European Commission leads on to broader questions about the general compatibility of long-term developmental and short-term security imperatives and about the extent of coordination necessary between different types of short-term crisis management operations.

Regarding the first case, the relations between military operations and external non-governmental actors involved in developmental assistance or emergency humanitarian aid – termed civil-military relations or CIMIC (Civil-Military Cooperation) in military jargon (see Vorhofer 2003) – have become both closer and more conflictual (see Duffield 2001). In some instances, the complexity of peace-building has led to the convergence of civilian and military tasks: the military, in addition to providing a protective security shield for relief and reconstruction efforts, has at times undertaken larger-scale infrastructure projects such as the construction of roads, refugee camps and hospitals. Also, and usually under exceptional circumstances, the military has provided humanitarian aid in situations where normal aid agencies could not operate (see Spence 2002; Heineman-Grüder and Pietz 2004). Since they are primarily trained for battle within a strictly hierarchical chain of command, these tasks come as novelties for many military organizations and their incorporation into military doctrine is a far from straightforward process. Some military organizations have been reluctant to take on new roles, pointing to the risks of decreased battle readiness and 'mission creep', the gradual shifting away of organizational activities from former mandates. On the other hand, civilian actors have criticized humanitarian actions performed by military agencies as both undermining the necessary neutral 'humanitarian space' of non-governmental relief organizations and as taking over their jobs (see Gourlay 2000; Lilly 2002). Coordination of relief work with military operations has so far predominantly taken place in the field and in an ad hoc manner, since progress towards strategic coherence at the European political level has generally been slow.

Similar problems of coordination arise among the different governmental actors engaged in the increasingly diverse formats of EU operations. In a departure from understandings prevalent in 'traditional' forms of peacekeeping that were based on physical interposition and the support of interim ceasefires, the new multidimensional and multi-actor 'peace support' operations are tasked with ensuring the constitution and perpetuation of a stable and sustainable peace (see Bellamy, Williams and Griffin 2004). To this end, the field of peacekeeping and external crisis management – formerly a monopoly of the military – incorporated a new set of actors drawn mainly from the police and judicial services. Today, the police and rule of law missions in charge of assisting with the establishment and training of democratic law enforcement and judicial services in the host nations make up a large part of EU crisis management operations. So far, however, the design of a coherent and integrated model for multi-agency peace support operations presents an ongoing challenge. The EU internal 'civil–military coordination' – CMCO in EU terminology – has been limited to calls for developing a 'culture of coordination' (cf. Council Document No. 14457/03) between the agencies. Integrated planning and operational capabilities are still a future prospect (see Dwan 2004: 15). In general, the example of EU crisis prevention and management policies showcased a series of problems arising out of the new complexity of the post-Cold War security agenda. Not only should security strategies be proactive in their timing, they also have to be comprehensive in scope. Hence, despite remarkable progress made in the overall institutionalization of the European security architecture, the transformation towards integrated and preventive security measures remains in its infancy.

Conclusion: the politics of risk

This chapter has illustrated that the risk paradigm poses a serious challenge to traditional security thinking since it avoids an exclusively military-based focus on security provision by taking into account the multilayered and complex nature of current security challenges. Yet, using the concept of risk as a prism through which to look at questions of security poses its own problems. The first fundamental challenge of understanding security provision in terms of risk management lies in the sociological notion of risk 'being what we make of it'. Risk perceptions and management strategies are subject to societal and political negotiations about what constitutes acceptable or unacceptable levels of risk. Security strategies thus depend on the weighing up and prioritization of specific 'risky' phenomena, often to the detriment of other issues. Here, the problem is that to an increasing extent, political decisions are taken under conditions of uncertainty or even ignorance about the nature of security risks and their harmful consequences. Thus, political steering is prone to subversion by unintended outcomes (Schelling 1978) and could even evoke precisely those risks it seeks to avert (Daase 2002a: 21).

Moreover, normative and legal difficulties of applying preventive principles to security policies loom large. In the international system of states, national security policies were based on the principles of reactive territorial defence, non-intervention into the domestic affairs of other sovereign states and the interdiction of wars of aggression. Preventive security policies have the potential to go against the grain of each of these principles by transferring 'the first line of defence ... abroad' (European Security Strategy: 7), as the example of preventive intervention has shown. In the domestic sphere, the challenge of implementing precautionary security measures whilst upholding democratic civil liberties seems even more difficult to solve. In particular, the case of counter-terrorism highlighted that a convergence of intelligence and domestic law enforcement functions towards 'preventive investigations' can prejudice the basic civil rights of the individual.

Ultimately, security policies dealing with risk can only master means and not ends. The pervasive uncertainty about the nature of security risks and about the likely effects of security strategies leaves political decision-makers bereft of the possibility of solving today's security challenges. Security strategists are left to choose among different means of reducing and managing specific risks, thereby taking a fundamentally political decision about the prioritization of certain security problems over others. In modern risk societies, the provision of security has moved from a perceived 'politics of necessity' during the Cold War to a 'politics of choice', which presents both chances and challenges.

Notes

1 Its fundamental proposition is that in cases of scientific uncertainty about the consequences of specific activities or products for health and environment, precautionary measures should be taken that reduce their possible harm, even if evidence of a causal link between the activity or product and the feared consequences is scientifically not conclusive.
2 See Falkenrath (2001) for a pre-9/11 perspective, Heng (2002) for a more recent point of view, and COM (2004)701 for a European perspective.
3 The other three risks were specified as terrorism, organized crime and the proliferation of weapons of mass destruction (European Security Strategy: 3f.)
4 Taken as a short-hand for intentionally as well as unintentionally faulty and undemocratic political structures, bad or weak governance is indicated by certain factors, which the Commission Communication on Conflict Prevention lists as 'poverty, economic stagnation, uneven distribution of resources, weak social structures, undemocratic governance, systematic discrimination, oppression of the rights of minorities, destabilising effects of refugee flows, ethnic antagonism, religious and cultural intolerance, social injustice and the proliferation of weapons of mass destruction' (COM (2001) 211).
5 'Characteristics of structural stability are sustainable economic development, democracy and respect for human rights, viable political structures and healthy environmental and social conditions, with the capacity to manage change without to resort to conflict.' (SEC (1996: 332); 'Treating the root

causes of conflict implies creating, restoring or consolidating structural stability in all its aspects' (COM (2001) 211: 10).
6 European Community actors include the Directorates General *External Relations* and *Development*, in particular the *Conflict Prevention and Crisis Management Unit* within DG RELEX, the implementing agency *EuropeAid* and the *European Humanitarian Aid Office* (ECHO). Examples of Country Strategy Papers can be found at http://europa.eu.int/comm/external_relations/sp/.
7 Committees and agencies specifically tasked with the development of crisis management procedures include the *Policy Planning and Early Warning Unit*, which reports to the High Representative, DG IX *Civilian Crisis Management and Coordination* and the *Police Unit* in the Council Secretariat, the *Council Committee for Civilian Aspects of Crisis Management* (CIVCOM), the *Military Committee* as well as the *Political and Security Committee*, which provides general political oversight.

References

Ackermann, Alice (2000) *Making Peace Prevail. Preventing Violent Conflict in Macedonia* (New York: Syracuse University Press).
Ackermann, Alice (2003) The Idea and Practice of Conflict Prevention, *Journal of Peace Research*, 40 (3): 339–347.
Ackermann, Alice and Antonio Pala (1996) From Peacekeeping to Preventive Deployment: A Study of the United Nations in the Former Yugoslav Republic of Macedonia, *European Security*, 5 (1): 83–97.
Adamson, Fiona (2002) *International Terrorism, Non-State Actors and the Logic of Transnational Mobilization: A Perspective from International Relations*. Online: Social Science Research Council.
Andreas, Peter and Richard Price (2001) From War Fighting to Crime Fighting: Transforming the American National Security State, *International Studies Review*, 3 (3): 31–52.
Baldwin, David A. (1997) The Concept of Security, *Review of International Studies*, 23 (1): 5–26.
Beck, Ulrich (1992) *Risk Society: Towards a New Modernity* (London: Sage).
Beck, Ulrich (1999) *World Risk Society* (Cambridge: Polity Press).
Beck, Ulrich (2002) The Terrorist Threat. World Risk Society Revisited, *Theory, Culture & Society*, 19 (4): 39–55.
Bellamy, Alex J., Paul Williams and Stuart Griffin (2004) *Understanding Peacekeeping* (Cambridge: Polity Press).
Bigo, Didier (2001) The Möbius Ribbon of Internal and External Security(ies), in Albert, Mathias; Jacobson, David and Josef Lapid (eds.) *Identities, Borders, Orders. Rethinking International Relations Theory* (Minneapolis: University of Minnesota Press), pp. 91–116.
Bothe, Michael (2003) Terrorism and the Legality of Pre-emptive Force, *European Journal of International Law*, 14 (2): 227–240.
Buzan, Barry, Ole Wæver and Jaap de Wilde (1998) *Security: A New Framework for Analysis* (London: Westview Press).
Coker, Christopher (2002) *Globalisation and Insecurity in the Twenty-first Century: NATO and the Management of Risk*, Adelphi Paper 345 (Oxford: International Institute for Strategic Studies).

Commission of the European Communities (1996) The EU and the Issue of Conflicts in Africa: Peace-building, Conflict prevention and Beyond. SEC (1996)332 final, 6 March 1996.

Commission of the European Communities (2001) Communication from the Commission on Conflict Prevention. COM (2001)211 final, 11 April 2001.

Commission of the European Communities (2004) Communication from the Commission on preparedness and consequence management in the fight against terrorism. COM (2004)701 final, 20 October 2004.

Council of the European Union (2003) Civil Military Co-ordination (CMCO). Council Document No. 14457/03, 7 November.

Council of the European Union (2003) European Security Strategy: A Security Europe in a Better World, 12 December.

Creveld, Martin van (1998) *The Sword and the Olive. A Critical History of the Israeli Defense Force* (New York: Public Affairs).

Cussen, Sarah (2004) *Background Paper on Existing Arrangements and Cooperation within the Framework of CFSP and ESDP*. Paper prepared for the Study Group on Europe's Security Capabilities, London.

Daase, Christopher (2002a) Internationale Risikopolitik: Ein Forschungsprogramm für den sicherheitspolitischen Paradigmenwechsel, in Christopher Daase, Susanne Feske and Ingo Peters (eds.) *Internationale Risikopolitik. Der Umgang mit neuen Gefahren in den internationalen Beziehungen* (Baden-Baden: Nomos), pp. 9–36.

Daase, Christopher (2002b) Terrorismus: Der Wandel von einer reaktiven zu einer proaktiven Sicherheitspolitik der USA nach dem 11. September 2001, in Christopher Daase, Susanne Feske and Ingo Peters (eds.) *Internationale Risikopolitik. Der Umgang mit neuen Gefahren in den internationalen Beziehungen* (Baden-Baden: Nomos), pp. 113–142.

Deighton, Anne (2002) The European Security and Defence Policy, *Journal of Common Market Studies*, 40 (4): 719–741.

Duffield, Mark (2001) *Global Governance and the New Wars. The Merging of Development and Security* (London: Zed Books).

Dwan, Renata (2004) *Civilian Tasks and Capabilities in EU Operations*. Paper prepared for the Study Group on Europe's Security Capabilities, Berlin.

Ericson, Richard V. and Kevin D. Haggerty (1997) *Policing the Risk Society* (Oxford: Clarendon Press).

Richard A. Falkenrath (2001) Problems of Preparedness. U.S. Readiness for a Domestic Terrorist Attack, *International Security*, 25 (4): 147–186.

Giddens, Anthony (1985) *The Nation-State and Violence* (Cambridge: Polity Press).

Gourlay, Catriona (2000) Partners Apart: Managing Civil–Military Cooperation in Humanitarian Interventions, *Disarmament Forum*, 3: 12–22.

Heineman-Grüder, Andreas and Tobias Pietz (2004) Zivil-militärische Intervention – Militärs als Entwicklungshelfer? In Christoph Weller, Ulrich Ratsch and Reinhard Mutz (eds.) *Friedensgutachten 2004* (Münster: Lit-Verlag), pp. 200–208.

Heng, Yee-Kuang (2002) Unravelling the 'War' on Terrorism: A Risk-Management Exercise in War Clothing? *Security Dialogue*, 33 (2): 227–242.

Hill, Christopher (2001) The EU's Capacity for Conflict Prevention, *European Foreign Affairs Review*, 6 (3): 315–333.

Hoffmann, Stanley (1995) The Politics and Ethics of Military Intervention, *Survival*, 37 (4): 29–51.
Houben, Marc (2003) *Better Safe than Sorry: Applying the Precautionary Principle to Issues of International Security*. Working Document No. 196 (Brussels: CEPS).
Howorth, Jolyon (2001) European Defence and the Changing Politics of the European Union: Hanging Together or Hanging Separately? *Journal of Common Market Studies*, 39 (4): 765–789.
International Crisis Group (2005) *EU Crisis Response Capability Revisited*. Europe Report No. 160 (Brussels: International Crisis Group).
Johnston, Les (1997) Policing Communities of Risk, in Peter Francis, Pamela Davies and Victor Jupp (eds.) *Policing Futures: The Police, Law Enforcement and the Twenty-First Century* (London: Macmillan), pp. 186–207.
Lilly, Damian (2002) *The Peacebuilding Dimension of Civil–Military Relations in Complex Emergencies: A Briefing Paper* (London: International Alert).
Luhmann, Niklas (1991) *Soziologie des Risikos* (Berlin: Walter de Gruyter).
Lund, Michael (1996) *Preventing Violent Conflicts: A Strategy for Preventive Diplomacy* (Washington: United States Institute of Peace Press).
Migdal, Joel S. (1988) *Strong Societies and Weak States: State–Society Relations and State Capabilities in the Third World* (Princeton, NJ: Princeton University Press).
Milliken, Jennifer (ed.) (2003) *State Failure, Collapse & Reconstruction* (Oxford: Blackwell).
Politi, Alessandro (1997) *European Security: The New Transnational Risks*, Chaillot Paper 29 (Paris: Institute for Security Studies of the Western European Union).
Prevention of Terrorism Act (2005), 11 March (London: HMSO).
Rasmussen, Mikkel Vedby (2004) 'It Sounds Like a Riddle': Security Studies, the War on Terror and Risk, *Millennium*, 33 (2): 381–395.
Scharpf, Fritz W. (1986) Policy Failure and Institutional Reform: Why Should Form Follow Function? *International Social Science Journal*, 108: 179–189.
Schelling, Thomas C. (1978) *Micromotives and Macrobehavior* (New York: W. W. Norton).
Schiff, Ze'ev (1999) Fifty Years of Israeli Security: The Central Role of the Defense System, *Middle East Journal*, 53 (3): 434–442.
Schnabel, Albrecht (2002) Post-Conflict Peacebuilding and Second-Generation Preventive Action, *International Peacekeeping*, 9 (2): 7–30.
Schneckener, Ulrich (2002) *Netzwerke des Terrors. Charakter und Strukturen des transnationalen Terrorismus*. Berlin: Stiftung Wissenschaft und Politik, SWP-Studie 42.
Sofaer, Abraham D. (2003) On the Necessity of Pre-emption, *European Journal of International Law*, 14 (2): 209–226.
Spence, Nick (2002) Civil–Military Cooperation in Complex Emergencies: More than a Field Application, *International Peacekeeping*, 9 (1): 165–171.
Strategy Unit (2005) *Investing in Prevention: An International Strategy to Manage Risks of Instability and Improve Crisis Response*. A Prime Minister's Strategy Unit Report to the Government (London: Cabinet Office).
Suhrke, Astri (1999) Human Security and the Interests of States, *Security Dialogue*, 30 (3): 265–276.
Tilly, Charles (ed.) (1975) *The Formation of National States in Western Europe* (Princeton, NJ: Princeton University Press).

Treverton, Gregory F. (2005) *Making Sense of Transnational Threats: Workshop Reports* (Santa Monica: RAND National Security Research Division).

Vorhofer, Peter (2003) Civil–Military Cooperation. Zur Evolution einer neuen Aufgabe in der Krisenbewältigung, *Österreichische Militärische Zeitschrift*, 6: 753–759.

Wæver, Ole, et al. (1993) *Identity, Migration and the New Security Agenda in Europe* (London: Frances Pinter).

Walt, Stephen M. (1991) The Renaissance of Security Studies, *Review of International Studies*, 35 (2): 211–239.

Walter, Christian et al. (eds.) (2004) *Terrorism as a Challenge for National and International Law: Security versus Liberty?* (Berlin: Springer).

Walzer, Michael (1977) *Just and Unjust Wars* (New York: Basic Books).

Wheeler, Nicholas J. (2000) *Saving Strangers. Humanitarian Intervention in International Society* (Oxford: Oxford University Press).

Wheeler, Nicholas J. and Ken Booth (1992) The Security Dilemma, John Baylis, and N. J. Rengger (eds.) *Dilemmas of World Politics. International Issues in a Changing World.* (Oxford: Clarendon Press), pp. 29–60.

Williams, Phil (2002) Transnational Crime and the State, in Robert Bruce Hall, and Thomas J. Biersteker (eds.) *The Emergence of Private Authority in Global Governance* (Cambridge: Cambridge University Press).

Wolfers, Arnold (1962) National Security as an Ambiguous Symbol, in Arnold Wolfers (ed.) *Discord and Collaboration. Essays on International Politics* (Baltimore, MD: Johns Hopkins University Press), pp. 147–165.

Zartman, William I. (ed.) (1995) *Collapsed States: The Disintegration and Restoration of Legitimate Authority* (Boulder, CO: Lynne Rienner).

7
Terrorist Attacks at Nuclear Facilities

Chandrika Nath

Introduction

The last 50 years have seen a global expansion of the nuclear power industry. There are 440 nuclear power plants in operation worldwide in 31 countries, generating over 16 per cent of the world's electricity. This has given rise to a wide range of facilities containing radioactive material: not only the nuclear reactors themselves, but also supporting activities, such as making fuel or storing radioactive waste.[1] In recent years there has been increasing public concern over the terrorist threat to such facilities, heightened by the attacks on the Twin Towers and Pentagon on 11 September 2001.

Any successful attack could lead to dispersal of significant amounts of radioactive material, contaminating large areas of land and exposing the general public to an increased risk of cancer, as well as causing widespread panic and disruption. The impacts of an attack would not be confined within national boundaries; successful sabotage of a nuclear facility in one country could have serious consequences overseas. Even an unsuccessful attack might affect public attitudes to nuclear power and therefore impact on the development of the nuclear industry worldwide.

The risks and consequences of terrorist attacks on nuclear facilities are a subject of longstanding public debate, involving a wide range of different actors. Governments, regulatory bodies, the nuclear industry and international watchdogs are involved in assessing risks and taking measures to address them. They release some information on these activities to the public, but much is classified. There is also widespread commentary by independent analysts and non-governmental organizations (for example, environmental groups). The issue receives considerable media attention, usually stimulated by the activities of these interest groups.

In this chapter I will discuss the different ways these actors influence the debate on nuclear terrorism. I will argue that it is difficult for members of the public to make an informed judgement on the level of risk they face, because much of the information needed to do so is classified. Moreover there is a vast amount of conflicting commentary on the issue, usually transmitted by the media. Thus there is considerable uncertainty and confusion surrounding the debate.[2]

In the first section I will explain how risk is defined, and how a different regulatory approach is needed to deal with the risk of deliberate attacks at nuclear facilities rather than the risk of nuclear accidents. In the second I will discuss how the risk of terrorist attacks is assessed and managed by the nuclear authorities, highlighting the role of international bodies such as the International Atomic Energy Agency. I will describe how these risks are communicated to the general public, comparing portrayals by industry, government, NGOs and the media. In the third section, I will discuss factors affecting public perception of this issue, as well as what is known about public attitudes, based on data from recent MORI surveys on public attitudes to nuclear power.[3]

How is risk defined?

The word risk is used in a variety of contexts. In general, it is used to define the harm associated with a given activity. Nuclear activities are potentially harmful because people might be exposed to 'ionizing radiation'.[4] This can cause a range of adverse health effects depending on the radiation dose received, from short-term acute radiation sickness, which is usually fatal, to an increased longer-term risk of developing cancer.

A commonly accepted definition of risk is that used by the UK's Health and Safety Executive,[5] which defines risk, and the closely related term hazard, as follows:

- risk is defined in terms of probabilities or frequencies, i.e. 'the chance of harm being realized';
- 'hazard' is defined as 'intrinsic propensity to cause harm, which would include the magnitude and type of harm as well as its potential to be realised'.

People are at risk of exposure to ionizing radiation as a result of accidents, as well as terrorist attacks. Although the consequences for the general public would be similar in both cases, different approaches need to be taken to managing these different risks.

Nuclear accidents

Nuclear plants are designed to cope with a range of different accidental scenarios known as design basis accidents. Accidents falling into this category are determined by:

- how often they are predicted to occur (their likelihood); and
- what their consequences would be (expressed in terms of the radiation dose received by a member of the public).

These factors are quantified using technical data and mathematical models. To obtain a licence to operate, operators of nuclear plants must demonstrate that for all design basis accidents the public would not be exposed to dangerous levels of radiation.

Within this framework, plants may not be designed to cope with certain very unlikely scenarios, even if their consequences would be extreme. There is no upper limit to the radiation dose people might receive if such accidents were to occur. The treatment of such low likelihood, high consequence accidents is a highly contentious issue.

Deliberate attack

It is not possible to quantify the intentions of a terrorist group. As pointed out by Leventhal and Alexander: 'Sabotage is not mathematically random and involves deliberate attempts to defeat safety systems.'[6] Thus, the accidental likelihoods just discussed are not applicable to deliberate acts. A different approach is required. Four distinct factors need to be taken into account to assess the risks and consequences of a terrorist act:

- Terrorist intentions: what terrorist groups are operating, and which, if any, have the intention to attack a nuclear facility and the resources to mount an attack?
- Vulnerability: how physically robust are nuclear facilities?
- Security: how effective are security arrangements at nuclear facilities?
- Consequences: what factors might affect the environmental and health impacts of an attack (e.g. weather conditions, emergency planning arrangements)?

Each of these factors must be considered by the nuclear authorities when deciding how to manage the risk of terrorist attack and is the subject of widespread public debate involving a range of different actors, as discussed below.

Note that this chapter focuses on the threat of sabotage in a nuclear plant. It does not consider the possibility that terrorists might attempt to steal material to make a nuclear weapon. There are different factors involved in assessing this risk, for example the feasibility of constructing and detonating a nuclear weapon. This would require a separate analysis.

How is risk assessed, managed and communicated?

Intelligence information

Debate over the nature of a terrorist threat to nuclear facilities occurs at two levels: the official level, to decide how to protect nuclear sites (most of these discussions are not made public); and in the public domain, with input from NGOs, the media and to a limited extent from the nuclear authorities. In the latter case the lack of information can lead to widespread speculation.

Official assessment of terrorist intentions

Decisions on what kinds of attacks on nuclear facilities require contingency plans are based on official intelligence information. The UN's nuclear watchdog, the International Atomic Energy Agency, has recommended that each country draws up a list of potential methods that terrorists might use to attack nuclear facilities, based on their own assessment of the threat. This list is referred to as the 'Design Basis Threat' (DBT). The contents of the DBT are secret, although some countries elect to make parts of it publicly available. Security arrangements at nuclear sites (discussed in the next section) are designed to cope with the scenarios specified in the DBT document.

Nuclear sites took measures to defend themselves against terrorist attack long before 11 September 2001. The threat was from different groups in different countries – for example in the UK, the main threat was considered to be from Irish Republican terrorist groups. In the former Soviet Union, the threat was largely from Chechen separatists. In the US, the Nuclear Regulatory Commission (NRC) took additional measures in the 1990s to protect US nuclear reactors against vehicle bomb attacks following the attack on the World Trade Center in 1993. Box 7.1 illustrates some of the scenarios included in the US DBT before 11 September 2001. However, after this date the threat to nuclear facilities worldwide was re-evaluated. Information published by the UK's nuclear security regulator (the Office of Civil Nuclear Security or OCNS) indicates that the current threat is largely considered to come from terrorist groups operating at the international level.

The changing nature of the terrorist threat is illustrated by changes to the DBT in the US, some of which have been made public. For example, the US DBT now assumes terrorists are well armed and equipped, trained in paramilitary and guerrilla warfare skills, willing to kill, risk death or commit suicide; and are capable of attacking without warning. Sites also have to defend against a larger number of terrorists than before, although what this number is, is not revealed. In the UK, OCNS mentions two possibilities which are now included in the DBT: a) that terrorists might use vehicles loaded with explosives to crash through perimeter defences, and b) that terrorists who were prepared to kill themselves or risk discovery might attempt to penetrate into [certain sensitive areas inside sites] to detonate explosives.[7]

Unofficial assessments and speculation

In spite of the wide range of measures taken to increase the protection of nuclear sites since 11 September 2001, there is little evidence in the public domain to confirm whether or not terrorists have attempted, or threatened to attempt, an attack on a nuclear facility. A statement issued by the National Commission on Terrorist Attacks upon the United States indicates that according to assertions 'reportedly made by various September 11th conspirators and captured Al Qaeda members whilst under interrogation', nuclear power plants were among the targets originally considered plans for the 9/11 attacks. However, these assertions were obtained at third hand, via written reports, rather than directly.

Nevertheless, public debate over the potential terrorist threat to nuclear facilities has greatly increased since 9/11. This is illustrated by a

---

Box 7.1 The Design Basis Threat in the US[8]

In the US, scenarios considered in the DBT before 11 September 2001 included:

- Attacks by well-trained and dedicated individuals, possibly with military training and skills.
- Attacks involving insider assistance.
- Attacks involving suitable weapons, up to and including hand held automatic weapons, equipped with silencers and having effective long-range accuracy.
- Attacks involving hand carried equipment, including incapacitating agents and explosives.
- Attacks involving a four-wheel land drive vehicle to transport attackers and their hand held equipment to the proximity of vital areas.
- An attack involving a four wheel drive land vehicle bomb and insider assistance (this was added following the 1993 attack on the World Trade Center).

vast increase in media coverage of the risk of terrorist attack at the Sellafield nuclear plant in Cumbria (see case study below). However, there was awareness of the threat to nuclear facilities even before these events. The possibility of aircraft attack, although not widely discussed, was raised by some analysts as far back as 1975, after media reports that terrorists had threatened to fly a plane into a US reactor.[9]

A range of potential modes of terrorist attack has been proposed in published literature, though they are not necessarily based on official information. Some of the scenarios put forward are considerably more extreme than those suggested by the nuclear authorities, for example:

- that terrorists might attempt to attack liquefied natural gas tankers as they pass coastal nuclear facilities, to bring about an explosion which would inflict severe structural damage on the facilities;
- the use of weapons from the air.[10] At a recent conference an independent analyst described nuclear facilities as 'pre-deployed radiological weapons for terrorists within the countries they would most like to hit'.[11]

In the absence of official information, one cannot say how realistic these scenarios are. Nevertheless, they generate considerable media coverage and thus can be assumed to play a key role in influencing public perception.

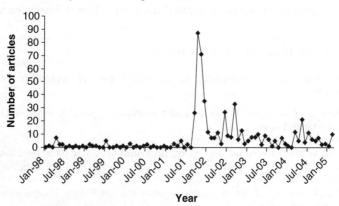

Figure 7.1 Keyword search for 'Sellafield' and 'Terrorism' in British print media from January 1998 to February 2005

A simple keyword search for references to Sellafield and terrorism in British national print media between 1998 and 2005 reveals that media coverage greatly increased after 9/11, with over 200 articles appearing between September 2002 and January 2003 (see Figure 7.1).[12] This is a long-term effect – although levels of coverage have declined, they still remain higher than before 9/11. However, other, smaller peaks and troughs in coverage can be seen over the five-year period. These are stimulated by a range of events, from protest group activities to the release of official reports. Not all of these events are security related, and in many cases they are nothing specifically to do with Sellafield. A closer look reveals that their focus has shifted towards discussing sabotage and aircraft impact rather than theft of material in recent years (see Table 7.1).

Security arrangements

Security regimes are in place at all nuclear sites to protect them against attack. Measures range from patrols by armed guards, to barriers and

Table 7.1 Events linked to media coverage of Sellafield

Peak	Subject of coverage
April 1998	Resignation of UKAEA Chief Constable; security exercises at Sellafield and Dounreay.
July 1999	First MOX (mixed oxide fuel) shipment from UK to Japan.
February 2000	Coverage of MOX data falsification story; debate over future of reprocessing at Sellafield.
May 2001	New Scientist article on the implications of theft of MOX by terrorists.
September 2001	Concerns over potential for terrorist attacks at Sellafield after events of 9/11 in US.
April 2002	Coverage of European Parliament Petition Committee meeting on discharges from Sellafield and Cap de la Hague after publication of a report commissioned by the European Parliament.[13]
June 2002	Coverage of planned MOX shipment from Japan to UK.
September 2002	MOX shipment approaches Britain, coverage of Greenpeace protests over shipment.
April 2003	Protest campaign over discharges from Sellafield.
May 2004	Parliamentary questions on alleged breaches of no-fly zones around nuclear plants.
July 2004	Release of Parliamentary Office of Science and Technology report on nuclear terrorism.
September/ October 2004	Shipment of plutonium from US to France.
February 2005	Reports of 'missing plutonium' at Sellafield.

vehicle checks. The operator of the nuclear site is responsible for making these arrangements. However, detecting and preventing attacks that are not mounted from the ground, such as aircraft attacks, are the responsibility of the national government. Although an attack in one country could have impacts overseas, responsibility for preventing it rests with the country itself. The adequacy of security arrangements at nuclear sites is a subject of heated debate.

International recommendations have been issued by the IAEA on how to protect nuclear facilities.[14] However, their effectiveness has been questioned by many analysts.[15] First, they are only guidelines and are not legally binding. Second, their language is often vague. To some extent such guidelines merely provide a 'lowest common denominator' so that countries can judge whether they are taking adequate steps to protect nuclear sites from terrorist attack. For example, the guidelines stipulate that sites should be protected by guards, but do not specify how many, or whether or not they should be armed. Policies on the use of force vary from country to country.

Indeed, the extent to which nuclear sites can actually be protected against terrorist attack by the use of force is a subject of heated debate. This is particularly marked in the debate over how to respond to attack by hijacked aircraft, where use of force against the aircraft would probably involve the death of innocent civilians. As pointed out by the House of Commons Defence Select Committee, 'establishing hostile intent would not be straightforward and whether to shoot [an aircraft] down would be a terrible decision to take. And there would be very little time in which to take it'.[16]

Actions taken by the authorities to protect nuclear sites often come in for a great deal of criticism. This is most evident where there are variations in international practice. For example, in France, following 9/11, anti-aircraft missiles were deployed outside the nuclear plant at Cap de la Hague.[17] The UK was criticized by some NGOs for not taking similar action.[18] This criticism was also reflected in media coverage: a search on the LexisNexis database identified eleven articles published in the UK national press, in the week after 9/11, which compared French and UK security policy and implied that the UK had taken too long to respond. Technical considerations, such as controversy over whether anti-aircraft missiles are actually a reliable form of defence at such facilities, were not discussed.

Security breaches at nuclear sites

In the UK, one of the main ways that security at nuclear sites is brought to public attention is when campaign groups try to gain entry into them. Greenpeace have successfully gained access to the Sizewell B nuclear plant in Suffolk on two occasions: 14 October 2002 and 13 January 2003. The second time, a group of over 30 protesters succeeded in scaling the perimeter fence. On both occasions the protesters gained access to the building housing the reactor itself, which should not have been possible. They also managed to scale the dome surrounding the reactor building.

A number of interesting observations can be made from these intrusions. There were differences in the way the two protests were covered by the media. The 2002 protest was primarily against suspected government plans for expansion of the nuclear industry. A LexisNexis search for Sizewell B revealed only nine articles about the protest in the UK national press between 1 October 2002 and 1 November 2002, with a length of 181 words. Of these, only four mentioned the aim of the protest. The others were short, light-hearted pieces mainly focusing on the fact that one protester was dressed as Homer Simpson. The aim of the second protest was to draw attention to security lapses. A search for coverage of the 2003 protest, between 1 January 2003 and 1 February 2003, revealed 22 articles with an average length of 331 words. Almost half of these specifically mentioned terrorism and the rest drew attention to security lapses. It seems likely that there was more media coverage of the second protest because the aims of the protest tied in with general media interest in discussing terrorism.

The second intrusion resulted in changes being made to site security – for example, razor wire was installed around certain parts of the site perimeter. The number of site inspectors was increased and security was reviewed at sites across the country. Similar changes were seen in 2002 when a *News of the World* journalist with fake references succeeded in gaining access to the inner area of Dungeness B nuclear power station in Kent, with a camera.[19] British Energy strengthened its searching and escorting arrangements after this event. However, while it revealed flaws in security arrangements, it is highly unlikely that one person alone could have inflicted serious damage to the reactor, due to the numerous safety systems in place.

Although activities by protest groups can highlight weaknesses in security arrangements, it would not be accurate to draw direct conclusions from them about the vulnerability of nuclear facilities to terrorist attack. Terrorists would operate very differently from groups of protestors; moreover protest activities are not handled as if they are genuine security threats. Following the Greenpeace incursions into Sizewell B the UK's Director of Civil Nuclear Security stated that he did not wish to 'divert effort and resources from maintaining counter-terrorist defences towards attempts to control and contain unarmed demonstrators' and pointed out that 'incursions by large numbers of demonstrators are wholly different in character and scale to attacks likely to be attempted by armed terrorists seeking access to sensitive sites'.[20] However, these considerations were not reflected in media coverage of the Greenpeace incursions.[21]

Physical robustness of nuclear facilities

Should terrorists succeed in overcoming security at a nuclear facility, the consequences of the attack would depend on how much physical damage was inflicted. This depends on the robustness of the facility. Many of the steps taken to provide protection against accidents will also provide some protection against terrorist attack. For example, the thick concrete walls around the cores of nuclear reactors, originally constructed to protect people from accidental exposure to ionizing radiation, also act as barriers to external attack.

However, many older nuclear facilities may not be able to withstand certain forms of attack such as commercial aircraft impact. This is a key cause for concern among independent analysts and NGOs.[22] At the time of the construction of these facilities (some as early as the 1950s), the chances of accidental aircraft impact were considered low enough not to be taken into account in plant design. The threat of deliberate aircraft impact was also considered inconceivable on the basis of intelligence information. There is widespread debate over this issue, and many investigations have been carried out into the structural damage likely to result from terrorist attacks (particularly aircraft impact) on older nuclear facilities. These analyses, which range from official assessments carried out by the nuclear authorities to unofficial assessments by NGOs and independent analysts, reach widely different conclusions, as illustrated by the example below.

Example: an aircraft attack at Sellafield nuclear plant

The Sellafield nuclear plant in Cumbria holds the largest inventory of radioactive material in Europe. There are several hundred facilities on the Sellafield site handling or storing radioactive material. Particular public concern has been focused on one specific facility, referred to as B215, where 1,000 tonnes of highly radioactive liquid waste are stored in tanks. In recent years the potential consequences of an aircraft impact on this facility have been a subject of heated debate over whether or not an aircraft could breach the walls of the tanks and release their hazardous contents into the environment.

- Industry reports: On the one hand, British Nuclear Fuels plc (the plant operator) have carried out 'detailed engineering assessments' which are 'confidential and not suitable to be placed in the public domain'. From these assessments BNFL says that it 'does not believe that the physical effects of an aircraft impact on this building would result in a loss of bulk shielding or containment'.[23] In other words, the public would not be exposed to harmful levels of radiation as a result of a terrorist attack on this facility. BNFL's conclusions must be taken on trust, since the details of these analyses, and of any assumptions made, are not released to the public for security reasons. Also, for security reasons, they cannot be subject to public peer review.
- Reports by NGOs and independent analysts: These, on the other hand, anticipate far more severe consequences for the general public. Such reports frequently cite an analysis carried out by Taylor in 1987.[24] This predicted that anything from several hundred thousand to several million people could be affected by a release of radioactive material from the B215 facility. However, these estimates are based on the assumption that 10 per cent of all the radioactive material would be released. This is an 'intelligent guess', which is not based on detailed calculations, because Taylor did not have access to detailed design information. BNFL claim these conclusions are 'unsubstantiated, entirely speculative and significantly exaggerate the consequences'.

Consequence management

Emergency response at national and international level

If terrorists did succeed in bringing about a release of radioactive material, the consequences for the general public would depend on what emergency measures were taken to protect them. People's exposure to radiation could be reduced by taking shelter, restricting food and water supplies, or if circumstances warranted, by evacuating people. In the UK, practice exercises, which take place regularly at local and also national level to test the emergency planning system, have revealed the potential for confusion and lack of co-ordination between the many agencies involved, particularly in communication with the media.[25]

The Chernobyl accident in 1986 (see next section) highlighted the importance of international communication in the event of an emergency at a nuclear facility. Tabletop practice exercises have taken place at the international level since the early 1990s. As with activities at the national level, these have highlighted the potential for confusion and lack of communication. They have shown that there is a need to use modern ways of transmitting data (e.g. the Web rather than telephone and fax) and for improvements in communication, both between officials in different countries and with the public and media.

At the international level, language poses another barrier. According to the Nuclear Energy Agency, English is the official language for international emergency communication, but many emergency management staff are unfamiliar with it and countries would need additional support with translation in an emergency.[26]

Chernobyl[27]

The Chernobyl accident in 1986 involved the largest release of radioactive material from a nuclear facility to date. For this reason, commentators frequently refer to Chernobyl when discussing the impact of a terrorist attack – for example, the phrase 'worse than 100 Chernobyls' is frequently used in the media to describe the potential impact of a terrorist attack on the Sellafield plant. However, there is still considerable uncertainty surrounding the health impact of the Chernobyl accident: estimates range from several thousand deaths from thyroid cancer, to over ten times that figure. This creates considerable confusion.

A UN report states that 'the nuclear industry acknowledges only very limited and closely defined consequences. On the other hand, some politicians, researchers and voluntary movement workers claim that the accident has had profound and diverse impacts on the health of many millions of people ... this uncertainty is a cause of widespread distress and misallocation of resources and needs to be addressed through rigorous and adequately funded international efforts.'

One of the key reasons for the uncertainty is that drawing data together from different countries is not straightforward. There are widespread variations in methodology and in many cases, incomplete information on how studies have been conducted.

What factors influence public perception?

In the previous section I showed that assessing and addressing the terrorist threat to nuclear facilities takes place in four stages: 1) assessing intelligence information; 2) implementing security measures at nuclear sites; 3) taking measures to increase the physical robustness of the facilities themselves; and 4) taking measures to manage the consequences of an attack. I have shown that for each factor there is considerable uncertainty because of restrictions on information, and different portrayals of the risk by different actors.

In this section I will discuss in more depth why it is difficult for a member of the public to make an informed judgement on the level of risk faced. I will discuss restrictions on information, make some general observations on risk assessments and comment on media coverage; all of which are likely to influence public attitudes. Finally, I will discuss data on public attitudes, from surveys carried out by MORI. These surveys suggest that the general public are less focused on the risk of terrorist attacks at nuclear facilities than the media.

Availability of information in the public domain

The nuclear authorities face an inevitable conflict between the need to keep the public informed about the risk of terrorist attacks on nuclear facilities and the need to protect sensitive information. For example, releasing detailed information on security arrangements at nuclear sites might allow a member of the public to make an informed judgement on the adequacy of these arrangements. However, such information might also assist a terrorist group planning an attack.

After the terrorist attacks on the Twin Towers and the Pentagon, a considerable amount of information on nuclear activities was withdrawn from official sources. For example, in the UK, maps showing the layout of nuclear sites were removed from official websites. Under the UK's Anti-Terrorism, Crime and Security Act 2001, it became a criminal offence for anyone to make a reckless disclosure of information that might affect the security of nuclear premises.

It is not possible for nuclear authorities to have complete control over all information placed in the public domain. Information can be obtained from other sources (e.g. NGOs), even if it is not made available by the authorities. Moreover, different countries have different policies on the availability of information, so even if there are tight restrictions in one country, information can be obtained from overseas via the Internet. For example, reports published in 2004 by the General Accounting Office in the US discuss in detail the design of flasks used to transport radioactive waste. Reports published by individual states even discuss in depth the kinds of artillery terrorists might use to pierce these flasks. UK authorities do not publish such detailed design information, but design has not changed substantially for several decades and technical information could probably be obtained from various sources. The vast amount of information available from these different sources is more likely to cause confusion than to help non-experts make informed judgements on the level of threat they face. To make an informed judgement, some form of discussion process is necessary. However, efforts towards increased transparency, which were being made in many countries including the UK, have been severely constrained in response to the events of 9/11.[28]

There are many examples of public hearings on nuclear issues prior to 9/11. For example in the UK, lengthy public hearings were held prior to the construction of the Sizewell B power plant, and the THORP facility at Sellafield. These hearings discussed a range of security and safety issues as well covering information about the design of the facilities themselves.

Although the reports from these hearings are still publicly available, it is highly unlikely that such public meetings would take place today, or that further reports with this level of detail would be published. Thus, it is not clear how informed public debate on construction of new nuclear power plants will proceed in future, although this is likely to be a highly topical issue in many countries, including the UK, where there is uncertainty over future energy supply and the construction of new nuclear plants has not been ruled out.

Some attempts at dialogue have taken place in recent years. For example, BNFL recently engaged in a dialogue with key parties about security issues. The aim of the exercise was to come up with recommendations for BNFL on how to improve security. The other bodies involved came from a wide range of different backgrounds – from official organizations such as the Office of Civil Nuclear Security to NGOs such as Friends of the Earth. The final report produced by these bodies reflects on the difficulties inherent in conducting a dialogue without full access to information: 'Inevitably the group had to operate without having unlimited access to the classified information that would have enabled it to conduct a comprehensive review of the existing security arrangements.' In particular, the evaluation was made without access to current intelligence on threats from adversaries – i.e. DBT. The report suggests that, in future, such an exercise could include 'selected citizens who have been security cleared to receive classified information'.

Risk assessments and their portrayal by the media

Risk assessments carried out by the nuclear authorities do not necessarily help the public to make informed judgements. This is because, in most cases, conclusions are provided, but full details of methodology are not made available. Moreover such assessments are not subject to public peer review because of their sensitivity. Thus, it is not possible for a member of the public to judge their reliability.

The assumptions made at each of the four stages of risk assessment outlined in the previous section are not always made clear, particularly when discussed by the media. In many cases, coverage can misrepresent information not only from official sources, but also from NGOs. For example, an article was published in the *Independent on Sunday* in May 2003 with the headline 'Attack on a nuclear plant could kill 3.5 million'. The first paragraph read: 'More than three and a half million people could be killed by a terrorist attack on a British nuclear plant, concluded a series of reports so alarming that even Greenpeace – which commissioned them – is unwilling to publish them.' However, the figures in this article are not drawn from any new studies by Greenpeace, but from the study carried out by Taylor in 1995 (see p. 121 above). The article used Taylor's 'worst-case scenario' figures, which assumed that no measures were taken to protect the public. Moreover, it suggests an atmosphere of secrecy, although the figures have been in the public domain for almost a decade. The *Independent on Sunday* article reflects a common trend. When risk assessments are reported, it is not always made clear what starting assumptions have been used, or even which analyses are being quoted.

The focus on scare stories and worst-case scenarios is in part a result of the fact that restrictions on information lead to speculation and citations of old and possibly outdated reports, in the absence of more recent information. However, it is reasonable to assume that fear of scare stories in the media influences decisions made by the nuclear authorities over what information to make public. Thus, there is a vicious circle.

The way that different interest groups express risk, and the language used, make comparison of different analyses very difficult. For example, a report commissioned by the US Nuclear Energy Institute in 2002 investigated the consequences of a ground-based terrorist attack on a US nuclear installation. The report concluded that 'the likelihood of one fatality' was 'less than 1 chance in 6,000 years'. A non-expert might have difficulty in interpreting this. According to the Committee to Bridge the Gap, a ground-based attack could 'result in sufficient radioactive to produce tens or hundreds of thousands of latent cancers'.[29] But because these two groups quantify the risk in different way, comparison of the two is difficult. Note also that the nuclear industry uses the word fatalities while NGOs use cancer or cancer deaths. The latter arouses more media interest, as is perhaps the intention.

Table 7.2 If you had to find out more about radioactive waste, how reliable and honest would you expect each of the following organizations to be? (expressed in terms of the per cent of the sample who believed that the group in question was extremely reliable/honest)

Group	2001	2004
Environmental campaign groups (e.g. Greenpeace)	38%	35%
University/academic scientists	28%	22%
The nuclear industry	28%	22%
The British Government	26%	19%

Whom do the public trust?

Which of the above actors has the most influence on public attitudes is likely to depend on the amount of trust the general public has in the different actors involved. *Eurobarometer* surveys indicate that public trust in charitable organizations and NGOs is generally higher than trust in government, big companies and the media, although attitudes vary in different member states.[30] A MORI survey into British attitudes to radioactive waste taken in 2004 revealed that the public believed that environmental/campaign groups were more honest and reliable than academics, the nuclear industry or government. However, the survey showed that confidence in all four bodies had declined since 2001 (see Table 7.2).[31]

What do surveys show?

A survey of 1,054 people in the UK carried out by MORI (commissioned by BNFL) in October 2004 revealed that 34 per cent (778) were opposed to the building of new nuclear power plants (30 per cent were in favour and the rest undecided). The 778 opposed to the building of new nuclear power plants were then asked their reasons. Although almost 25 per cent were opposed to new nuclear power because they regarded nuclear plants as 'unsafe/high risk', the principal concern was not terrorism. Accidents came first (16 per cent), followed by waste (10 per cent) and radiation (9 per cent). Only 1 per cent said that they were concerned about terrorism. In comparison, 7 per cent were concerned about terrorism in 2003, and 2 per cent in 2002. Thus there appears to be no clear trend in attitudes towards terrorism, but it is less of a concern than accidents or radioactive waste, according to this survey.[32]

There is slightly more concern over terrorism among politicians. A survey of 48 Members of Parliament opposed to new nuclear power plants revealed that 16 per cent were concerned over terrorism. However, this was still less of a concern than safety (55 per cent).[33]

The MORI surveys indicate that the risk of terrorist attacks at nuclear facilities is not a major fear for the general public. However, it is not possible to draw general conclusions from these limited data. We have not carried out detailed surveys of public attitudes to nuclear power as part of this study; they could be the subject of future work.

Conclusions

Faced with a vast amount of published commentary, usually transmitted via the media, it is difficult for the public to make an informed judgement on the level of risk presented by the possibility of terrorist attacks on nuclear facilities. There are many reasons for this. Assessing the risks of terrorist attack is an inherently uncertain exercise: it relies on many unquantifiable factors, such as terrorist motives and intentions. This, coupled with restrictions on information, means that assessments carried out by different groups often reach widely different conclusions. Also, because most information in the public domain comes from groups with either a pro- or anti-nuclear stance, the public are often exposed to conflicting reports.

A study of UK media coverage indicates that there has been a marked increase in coverage of the risk of terrorist attacks on nuclear facilities, since the New York and Washington DC incidents on 11 September 2001. There is a tendency towards scare stories and reporting of worst-case scenarios. This is partly a result of the limited information available in the public domain, which leads to speculation rather than evidence-based debate.

Governments face a difficult balance between keeping the public informed and the need to keep sensitive information out of the hands of terrorists. Any future public dialogue on new nuclear power plants might be severely constrained by the consequent lack of publicly available information.

We have not carried out detailed surveys of public attitudes to nuclear power as part of this study; they could be the subject of future work. However, surveys carried out by MORI indicate that fear of terrorist attacks at nuclear facilities are not one of the major reasons people cite if they are opposed to nuclear power, in spite of the extensive media coverage.

Nuclear facilities are not the only ones at risk of terrorist attack. There is a wide range of potential targets – for example, utilities, chemical plants and petroleum refineries. Some of these have lower levels of security than nuclear facilities. An informed debate on the risk to the general public from terrorist attacks on nuclear facilities would need to consider this. However, there is no established framework for comparing these different risks.

Notes

1. All these facilities are referred to hereafter as nuclear facilities.
2. This chapter builds on a report produced by the author for the UK Parliamentary Office of Science and Technology in July 2004, entitled 'Assessing the risks and consequences of terrorist attacks on nuclear facilities'. This was a report produced for UK parliamentarians. It aimed to provide a balanced and impartial overview of the issue, based on publicly available information. See http://www.parliament.uk/documents/upload/POSTpr222.pdf
3. MORI (Marketing Opinion and Research International) is an established market and public opinion research agency which carries out work for a range of public and private sector bodies. See www.mori.com.
4. 'Ionizing radiation' is radiation emitted by radioactive materials and can damage living tissue as it passes through it.
5. The UK's safety regulator.
6. Paul Leventhal, *Preventing Nuclear Terrorism*, 1987.
7. Office for Civil Nuclear Security, *The State of Security in the Civil Nuclear Industry*, April 2002–March 2003 (paragraph 35).
8. US General Accounting Office (GAO), *DOE Needs to Resolve Significant Issues Before it Fully Meets the New Design Basis Threat*, GAO-04-623, April 2004; *Several Issues could impede the Ability of DOE Office of Energy, Science and Environment to meet the May 2003 Design Basis Threat*, GAO-04-894T, June 2004. The US GAO is the audit, evaluation, and investigative arm of Congress.
9. David Krieger, Terrorists and Nuclear Technology, *Bulletin of the Atomic Scientists*, 31, 1975.
10. *Robust Storage of Spent Nuclear Fuel*, Institute for Resource and Security Studies, January 2003.
11. 'Nuclear analyst highlights risks of terror attack at nuclear installations', press release by the Nuclear Free Local Authorities, 7 March 2005.
12. This keyword search was carried out using the 'LexisNexis' search engine. Although the exact number of articles depends on the specific keywords used and may vary for different search engines, the LexisNexis search can be used to illustrate general trends.
13. *Possible Toxic Effects from the Reprocessing Plants at Sellafield (UK) and Cap de la Hague (France)*, a study prepared by WISE-Paris within the STOA (Scientific and Technological Options Assessment) programme of the European Parliament, November 2001.

14 International Atomic Energy Agency, *The Physical Protection of Nuclear Material and Facilities*, Information Circular 225, 4th revision. www.iaea.org/worldatom/Programmes/Protection/inf225rev4/rev4_content.html
15 See, for example, Bunn and Steinhausler, Guarding Nuclear Reactors from Terrorists and Thieves, *Arms Control Today*, October 2001. www.armscontrol.org/act/2001_10/bunnoct01.asp.
16 Defence Select Committee, 6th report, session 2001–2002, *Defence and Security in the UK*, July 2002.
17 These were removed within three weeks; however, anti-aircraft missiles are designed to be mobile and rapidly deployed.
18 Greenpeace, *The Potential for Terrorist Strikes on Nuclear Installations*, 2001.
19 *News of the World*, Nuke Nightmare, 8 September 2002.
20 Office for Civil Nuclear Security, *The State of Security in the Civil Nuclear Industry*, April 2003–March 2004.
21 See, for example, 'Showing Saddam the Way', *Daily Star*, 15 January 2003; 'Yesterday protesters penetrated the heart of this nuclear plant. What if it had been Al Qaida', *Daily Mail*, 14 January 2003.
22 See, for example, John H. Large, *The Implications of September 11th for the Nuclear Industry*, Large and Associates, Consulting Engineers, London and Mycle Schneider, World Information Service on Energy (WISE) Paris, France, February 2003.
23 Personal communication from BNFL, July 2004.
24 Peter Taylor, *Consequence Analysis of a Catastrophic Failure of Highly Active Liquid Waste Tanks serving the THORP and MAGNOX Nuclear Fuel Reprocessing Plants at Sellafield*, Nuclear Policy and Information Unit, Manchester Town Hall, Manchester, February 1994.
25 For example, in a practice exercise simulating an accident at the Bradwell nuclear power station in May 2002, there was even confusion about power station at which the simulated accident had occurred. See Emergency Planning Services, BNFL Magnox Generation, Bradwell Level 3 Extendibility Exercise, 10 May 2002.
26 Nuclear Energy Agency, Organisation for Economic Co-operation and Development, *Experience from International Nuclear Emergency Exercises*. See http://www.nea.fr/html/rp/reports/2001/nea3138-INEX2.pdf .
27 The Chernobyl disaster was the largest release of radioactive material to date from a commercial facility. It has led to the resettlement of about 370,000 people in the worst affected areas (parts of Russia, Ukraine and Belarus) and to widespread agricultural restrictions – around 150,000 km^2 of land (the combined area of England and Wales) was still defined as contaminated in the year 2000 (according to UNSCEAR). Levels of radioactivity were high enough to warrant countermeasures (mostly food-related) in most European countries.
28 Note, however, that there are some steps taken to keep the public informed – for example, in the UK, the nuclear security regulator publishes an annual report on its activities.
29 Committee to Bridge the Gap is a US-based nuclear watchdog group that provides technical, legal, and organizing assistance to communities near existing or proposed nuclear projects.

30 *Eurobarometer* surveys are public opinion surveys carried out by the European Commission in EU member states. See *Standard Eurobarometer*, http://europa.eu.int/comm/public_opinion/archives/eb/eb62/eb62_en.htm
31 Radioactive Waste Survey, Final Report, Research Study conducted for Nirex by MORI, March 2004. http://www.nirex.co.uk/foi/nxconsult/radwastesurvey_march2003.pdf .
32 Radioactive Waste Survey, Final Report, Research Study conducted for Nirex by MORI, March 2004.
33 Personal communication, MORI, March 2005.

8
The Case of Oil Exploitation in Nigeria

E. Remi Aiyede

The political economy of oil exploitation in Nigeria

Nigeria was under firm military autocracy and absolutism for close to 29 years after 1966 when the military made their first incursion into the country's government and politics, following the collapse of the first republic. Authoritarian governments were interrupted only by a brief period of civilian rule in the second republic (1979–83) before the return to civil rule in 1999. Indeed, Nigeria's march to constitutional democracy has been a chequered one, marked by anti-colonial struggles, crises, coups, counter-coups and a thirty-month agonizing civil war between 1967 and 1970. Nigeria can be said to be in its sixth political phase in its history as a single political entity, namely:

- The era of colonial autocracy and absolutism, that is, the period of formal colonialism until 1 October, 1960 when the country gained independence.
- The emergence of constitutional democracy between 1960 and 1966.
- The return of military autocracy and absolutism (1966–79).
- The restoration of constitutional democracy (1979–83).
- The second period of military autocracy and absolutism (1983–99).
- The return to constitutional democracy, 29 May 1999 to the present.

Nigeria is composed of over 250 ethno-linguistic groups with three dominant ethnic groupings: Hausa/Fulani in the north, Yoruba in the west and Igbo in the south-east. Naturally, these majority ethnic groups became the dominant players in politics once the British began to grant self-rule to the three regions at independence and have remained so. The untoward implication of these for minority peoples and communities is very clear from an exploration of the use to which state resources have been put in a contest in which strategic positions of the state (military, political offices, bureaucracy) are filled increasingly, especially from the 1979, according to regional and proportional representation.

Three critical areas are worthy of note. The first is privatization of public office. Politics in Nigeria has been described variously as prebendal, neo-patrimonial, clientelistic or predatory, which means that 'the existing offices of the state may competed for and then utilised for the personal benefit of office holders as well as their reference or support group' (Joseph, 1987). The second is that majoritarian politics is reflected in policy choices that do not favour the minority oil-producing communities, as is demonstrated in the changes in the revenue allocation formula. Since oil became the major source of revenue, derivation has increasingly become insignificant as a factor of allocation of revenue. The third is the consequential marginalization of oil-producing communities who are largely minorities in terms of development projects and revenue allocation in the struggles between dominant majority groups for power, as the former lost control over oil resources. Thus, the oil communities have had to engage the Nigerian state in an epic but alarming struggle for justice and equity (Joseph 1987; Graf 1988; Obi 1995; Lewis 1997; Aiyede 2002).

To appreciate the significance of oil exploitation to politics and social order in Nigeria, it is important to understand how oil exploitation has become central to resource allocation and distribution in Nigeria.

At independence Nigeria's economy and public revenue were largely derived from the export of agricultural produce such as cocoa, cotton, rubber and groundnuts. As the mainstay of the economy, agricultural produce accounted for 64.1 per cent of national output in 1960. Until 1966 oil accounted for less than 15 per cent of national output even though oil had become the fastest growing sector of the Nigerian economy. In 1970 oil revenue accounted for 58.01 per cent of total exports. Following a growth rate of 13 per cent between 1986 and 1992 it began to account for over 90 per cent of foreign exchange, 70 per cent of budgetary revenue and 25 per cent of Gross Domestic Product (GDP). At the same time annual production of Nigeria's cash crops fell by 43, 29, 65 and 64 per cent respectively between 1970 and 1982. This situation shows quite clearly that the inflow of oil revenues was not used to provide impetus for growth in other sectors. Indeed, the share of agriculture in GDP declined to 37 per cent in 1991 and 41 per cent in 1996 in spite of a series of programmes, e.g. the green revolution, back to land and Structural Adjustment Programmes, introduced by the government from the 1970s to boost agricultural production. Thus, the entire economy has remained heavily dependent on oil.

The expansion of the oil sector was so rapid, massive and dramatic that it transformed the politics, economy and the character of the Nigerian state. It provided immense benefit, especially in the form of increased income to the state. For instance, oil revenue accruing to the Nigerian state increased from $724 million in 1970 to $24,933 million in 1980, that is from 58.01 per cent of total export to 96.14 per cent in 1980 (Khan 1994: 184). On the other hand, it transformed the Nigerian economy into a mono-mineral economy, the state into a *rentier* state and the population into consumers rather than producers. In the last case, the productive sectors of the economy have become overshadowed by massive increase in imported goods, both capital and consumer, due to the availability of cheap income from petroleum. Table 8.1 shows the situation as Nigeria enters the twenty-first century.

The dominance of oil resources and the stultification of the productive sector of the economy had made politics the most lucrative business in town. Thus, not only has the struggle for power, with its promise of control of oil revenue, intensified political conflicts, but the availability of oil rents has helped sustain personalized military regimes against popular pressure. Oil revenues underwrite the various measures and

Table 8.1 Oil in the revenue profile of Nigeria

	$ Million			₦ Million		
Year	Total Export	Oil Export	per cent	Total Revenue	Oil Revenue	per cent
1988	6,931.7	6,319.0	91.6	27,595.0	19,831.7	71.86
1989	7,870.9	7,469.8	94.9	47,798.3	39,130.5	81.86
1990	13,671.1	13,265.6	97.03	69,788.2	65,215.9	93.44
1991	12,264.3	11,792.3	96.15	100,991.6	82,666.4	81.85
1992	11,886.11	11,641.7	97.94	190,453.2	164,078.1	86.15
1993	9,924.4	9,696.6	97.7	192,769.4	162,078.1	84.09
1994	9,415.1	9,170.7	97.4	201,910.8	160,192.4	79.33
1995	10,635.8	10,350.1	97.31	459,987.5	279,902.3	60.85
1996	16,153.6	15,866.4	98.22	520,190.0	408,783.0	78.58
1997	15,207.30	14,850.10	97.65	582,811.1	416,811.1	71.51
1998	8,971.20	8,564.70	95.46	463,608.8	324,311.2	69.95
1999	12,876.00	12,664.90	98.36	949,187.9	724,422.5	76.32
2000	19,441.40	18,897.20	97.2	1,906,159.7	1,591,675.8	83.5
2001	18,927.40	18,677.10	98.64	2,231,532.9	1,707,512.8	76.51

Source: *Central Bank of Nigeria Annual Report and Statement of Account* (various years).

strategies of regime sustenance. One important aspect of these measures is territorial fragmentation: the creation of new states and localities by the government – hence, from a four-region federation in 1963 to a twelve-state structure in 1967, to 19 in 1976, 21 in 1987, 30 in 1991 and 36 states from 1996. The number of local governments has expanded from 301 in 1979 when the central government intervened in the local government system to 774 in 1996.

In 1963, Nigeria embarked on a state-creation exercise to allay fears of domination by minority communities and interests. It was thought that creating a state for minority groups would protect them in regions dominated by major ethnic groups by expanding the scope for self-determination. The second reason (particularly relevant to the 1967 exercise) was to restructure the federation in such a way that no one state or group of states could threaten the corporate existence of the country or hold the country to ransom. In carrying out the exercise, particular consideration was to be given to economic viability (Gowon 1996: 27). As time passed, however, economic reasons became completely overshadowed by political expediency. For instance, the 1987 and 1991 exercises were informed by three principles according to the views of the prime actors in the process. The first was social justice, which implied the mollification of those who had lost out in previous state-creation exercises. The second was the principle of an even spread of development centres across the country, since new states and localities were to have direct access to federal funding (i.e. oil revenue). The third was equal distribution of units between the north and south, and east and west. In doing so, recognition was to be given to historical, sociocultural and geographical relationships among the country's over 250 ethnic groups (Suberu 1994: 16, Aikhomu 1996: 53).

The bottom line is that these exercises have reduced states and local governments to mere channels for the distribution of revenue from centralized sources, albeit inequitably, to the constituent communities. In doing this, it removed the independent revenue sources of sub-national governments, weakened administrative effectiveness and undermined accountability of the lower tiers of government to their populations. Thus, Nigerian federalism has been essentially distributive in that the more populous and larger communities take more revenue from the coffers of government even though oil deposits are located in minority communities.

Transnational risks, actors and their strategies

The greater exploitation of oil than other natural resources carries with it a lot of dangers and uncertainties that carry both ecological and social dimensions. Some of these – such as pollution and gas flaring – know no boundaries and hold great uncertainties for sustainable development for people as well as the environments where the exploitation takes place. Where precautionary measures are not taken, they inevitably generate social dislocations arising from forced physical resettlement. Sudden and dangerous pipe leakages or bursts resulting in oil spillages endanger the health of human, plant and aquatic life. Thus, risks arising from oil exploitation have provoked sharp reactions from communities where such exploitation is carried out, leading to severe and protracted conflicts between the Nigerian state and those communities, between those communities and oil corporations, and inter-communal strife over compensation, land ownership and questions relating to domination and social justice. Indeed, the calculations by these stakeholders of their relative advantage or disadvantage and the strategies adopted in pursuing their 'divergent' interests hold great implications for global governance and sustainable development.

Risk management has become one of the most critical elements of oil exploitation in the Niger Delta region. Risk management is about choices; which risks are worthy of consideration, which remedy to choose, how to trade off competing risks, balance competing factors. Risk is a subjective concept and reflects power relations in its analysis and management. There is, therefore, a need for an effective global framework for risk management that incorporates global inequality in terms of power, influence, resources and technical capacity for projecting risk among actors in natural resources exploitation (Risse-Kappen, 1995; Fone and Young, 2000; Pearce, 2000). From the perspective of the disadvantaged actors, global civil society holds great potential for social justice (Edwards, Hulme and Wallace, 1999). This chapter explores global civil society engagement with grassroots movement for environmental and social justice in Nigeria's petro-dollar economy, as a part of the challenge of transnational risk management.

The following sections analyse the interests, aspirations and strategies adopted to pursue the separate interests of the major actors in the exploitation of oil in Nigeria.

The Nigerian state

Oil has become the single most important material base of the Nigerian state, accounting for over 70 per cent of total revenue and 76 per cent of total export. In the postwar period, oil exploitation has become the live wire of the state's activities: bureaucracy, defence, infrastructure and social services. The implication is that any threat to oil exploitation is simultaneously a threat to the state. Being particularly subject to the vagaries of the international oil market as expressed in the balance of payment crisis that it continues to suffer since the oil glut of the early 1980s, the state is committed to ensuring that local efforts to disrupt oil production is stopped with all the powers at its disposal.

The strategic importance of oil first became apparent during the civil war. When the Igbos declared independence for Biafra, the new country included the minority oil-producing communities even though the consent of these non-Igbo minority communities weas never sought before the secession was declared. The Nigerian government, which had hitherto turned a deaf ear to cries of domination by these ethnic minorities, quickly divided the country into twelve states, with the minority oil-producing states divided between two, Rivers and Bendel. This raised the hope of autonomy for these minority communities within the Nigerian federal structure and the communities quickly joined the Nigerian effort to quell the secession attempt by Biafra (Adejumobi and Aderemi 2002).

However, state creation, which was successful in allaying the fears of these minority groups at the inception of the civil war in 1967, became an instrument for their further marginalization by successive military rulers when political considerations came to dominate as the basis for the creation of additional states.

While the strategy of safeguarding oil production has involved the use of both carrot and stick, force has dominated the action of the government in practice. For instance, the petroleum sector was one of the sectors considered to be an essential service sector when strikes were outlawed in the 1970s. Strikes by oil workers were dealt with decisively.

Furthermore, efforts by oil communities in pursuit of environmental and social justice have been met with violence by successive military regimes. For instance, the high mobilization and co-ordinated mass movement for environmental justice by the Ogonis were met by intense repression. The Ogoni region was militarized and the military special Task Force under Major Paul Okutimo terrorized the Ogoni region between 1992 and 1998. In 1995, after a kangaroo trial between 1994 and 1995, nine leaders of the Movement for the Survival of the Ogonis (MOSOP), including its leader, Ken Saro-Wiwa, were hanged in November 1995. Saro-Wiwa's appeal was still pending before the Court of Appeal and the Federal High Court when he was executed. As a result of this the country was suspended from membership of the Commonwealth. (Membership was restored only after the return to democratic constitutional rule in 1999.)

The same mindless brutality was unleashed in 1999 when Odi, an oil-producing community in the Niger Delta, was razed by soldiers on the orders of President Obasanjo in reaction to the kidnap and killing of seven policemen by armed youths there. Scores of innocent people were killed; thousands were rendered homeless, made to refugees and displaced persons in their own communities. This incident has remained a blot on the human rights record of the current government. The national government has also visibly supported the use of mercenaries by oil corporations in defence of their installations, an action that was challenged by the Delta State government in October 2003.

Apart from the strong-arm tactics, the national government, under pressure from the international human rights regime, the increasingly violent agitations by youths in the oil-producing communities, multilateral organizations (the United Nations, Commonwealth) and the global civil society movement, has set up structures to respond positively to the needs of the Niger Delta peoples. In 1988 the Federal Environmental Protection Agency (FEPA) was established to deal with environmental problems in oil and other sectors. In 1993 the Oil Mineral Producing Areas Development Commission (OMPADEC) was created to provide infrastructure and social amenities as part of the effort to respond specifically to the development needs of the Niger Delta region after several years of claims that the terrain there was too expensive and difficult to develop. Several laws and regulations were then introduced to prevent spillages and gas flaring by imposing penalties for such occurrences on the oil companies. Then, in 2000, the Niger Delta Development Commission (NDDC) was set up to replace OMPADEC due to the failure of the latter to achieve the objectives for which it was set up (Osaghae 1995; Obi 1997; Ikporukpo 2002).

The transnational oil corporations

The multinational oil corporations are the most significant agencies of oil exploitation. They have the capacity to explore and determine where oil deposits exist in commercial quantities. They have the technical expertise and the capital to exploit and refine oil for commercial use. The terms on which a foreign multinational enters a country to explore for and produce its natural resource reflects largely only the careful balance of the estimated risks and rewards.

In fact, until the oil companies' arrival, the oil-producing communities have lived on their lands without even knowing that such deposits exist and have no inkling of the possibilities of wealth and its potential for environmental degradation.

Furthermore, oil companies invest huge capital and expertise in the joint ventures that carry out oil exploitation in Nigeria. The volume of shares held by these companies run into billions of US dollars. These investments do not only have to be secured, the operations of the companies must be such that they ensure returns on these investments. Hence operational control is usually vested in these corporations reflecting their centrality above the state in the business of oil. Apart from the risks of investing huge sums in oil exploration, the fluctuations in the international oil market and operational accidents, the oil companies have had to contend with the disruption of production arising from the activities of protesting communities from the early 1990s. Thus, dealing with these pressures from the communities has been a major preoccupation.

Initially, the oil companies resisted the demands of these communities, maintaining that they were illegitimate. Shell Petroleum Development Company (SPDC), for instance, in a statement issued in 1993, argued that 'the most important contribution the company can make to the social and material progress of Nigeria is performing efficiently its direct line of business ... [the company cannot] pre-empt the responsibilities of the ... government in providing and maintaining social amenities and services' (see Ojo 2002: 39)

To curb the activities of these communities and their organizations, like the Movement for the Survival of Ogoni People (MOSOP), the transnational oil companies maintain strong security outfits to protect their installations, flow stations and employees. This is backed by strategies such as silence, defiance or denials in the face of accusations of irresponsible behaviour and reports of dangerous incidences of oil spillages. In using silence and denials, the companies 'hide problems, pretend that problems do not exist or insist they are imagined'. The corporations may also defy the communities by issuing statements indicating that they will not alter policy positions because of protest or court action, persisting on a particular course of action regardless of mounting protests.

Oil companies have also incorporated leaders of these communities, state official and youth leaders by paying them, appointing them to decision-making structures in the oil companies, awarding lucrative contracts to them and having the cost of holidays and medical treatment in the mother countries of the oil companies paid for. Many of those bought over are silenced or act in ways that compromise the interests of their own people. While these were effective until the 1980s, the high mobilization among the masses has led to these leaders being branded 'vultures' (traitors). The anger unleashed on some of these compromised leaders accounts for the killing of the Ogoni Four, which culminated in the execution of Saro-Wiwa, thus occasioning division among the ranks of the oil-producing communities. Ultimately, the companies and the Nigerian state deepen these divisions, exploiting them to divert protest. Currently, several host communities are up in arms against each other over the unequal distribution of community development projects (see Iyayi 2000: 159–67).

Ultimately, the protests became internationalized and oil companies increasingly face pressures from the international environmental movement, the United Nations and the Commonwealth, especially after the death of Saro-Wiwa in 1996. The companies have had adopt a more humane approach, taking into consideration the new international human rights regime, and the risk arising from social issues protests.

The oil-producing communities

When oil exploration began in 1958 in Nigeria, indigenes of the oil production communities received the development with joy. Many believed that oil would transform their lives for the better. They had expected that a huge percentage of the revenue from oil would return to develop their communities, and looked forward to a more robust livelihood.

However, by the late 1960s it had become clear that their hopes for a better life were misplaced. Slowly but relentlessly oil exploration and exploitation activities generated serious hazards such as gas flaring, oil spillage, indiscriminate construction of canals and waste dumping. The oil multinationals acquired lands from the Nigerian governments and pipelines were laid across community farmlands thanks to the Land Use Decree of 1973 without compensation to the communities. Neither were fiscal and social infrastructures provided for the communities. By the mid-1980s when the Nigerian government introduced structural reforms of the economy the level of degradation and deprivation in the oil-producing areas had reached provocative heights. But the government under severe constraints by its creditors to curb social spending ignored the complaints of these communities. This attitude had started in the early 1970s when Ogoni kings and community leaders wrote to the government about 'Damages done to life line by the continued presence of Shell BP Company of Nigeria, her installations and exploration of crude oil on our soil' (Okonta and Douglas 2001).

By 1990, with worsening environmental degradation and in the face of significant infrastructural developments in Lagos, Abuja and major non-oil-producing state capitals, coupled with the comparative luxury, leisure and affluence in which oil workers live, it dawned on these communities that they had become victims of a conspiracy involving the Nigerian state and oil multinationals to utilize the resources in their land at the expense of their survival. Their protest against the Nigerian state and the oil companies exploded into a full-fledged movement for a better deal. The government's initial reaction was to ignore them. Later it armed itself with laws and policies (of central control and ownership of natural resources) and intensified the use of security forces to secure oil installations. But this served to heighten the struggle of those minority oil-producing communities and radicalize their demands, supported by a level of mass mobilization hitherto unknown in the country, especially with the formation of the Movement for the Survival of the Ogoni people (MOSOP) in 1990 under the leadership of Ken Saro-Wiwa. The movement for environmental and social justice in the Niger Delta region since then has expanded in terms of its organizational expression. A brief list includes: the Movement for Reparation of Ogbaia (MORETO), the Movement for the Survival of Ijaw Ethnic Nationality in the Niger Delta (MOSIEND), the Ijaw National Congress (INC), Urhobo Development Forum and the Pan-Niger Delta Resistant Movement (CHIKOKO) (Douglas 1998).

Many of these groups have published and presented their demands to the Nigerian state and the oil corporations in well-articulated documents. Popular among these are the Ogoni Bill of Rights (26 August 1991), the Kaiama Declaration (December 1998), the charter of demands of the Movement for the Survival of the Izon (Ijaw) Ethnic Nationality in the Niger Delta (MOSIEND) (October 1992) and Niger Delta Resistant Movement (CHIKOKO) (August 1997). Several other organizations, subgroups of these larger and more popular organizations, have also articulated their demands. Most of these demands include:

- Compensation for oil-related environmental degradation.
- Political autonomy or the right to self-determination as a distinct and separate entity outside Nigeria.
- The immediate end to environmentally damaging economic activities by transnational corporations.
- Abrogation of the Land Use Decree and the Petroleum Decree, which rob Niger Delta communities of their right to control their land and mineral resources.
- The immediate withdrawal from the communities of all military forces of occupation and repression by the Nigerian state.
- That the Nigeria federation be run on the basis of equality and social justice.
- Employment in the oil industry and provision of basic social amenities (development projects) for their communities.

These demands have emerged as the sense of being victims became widespread among the indigenes of these oil-producing communities (Osaghae 1995).

In pursuit of these demands the communities and their organizations have made representations to the government and the oil companies. Considering the grave risks the specific actions such as oil spillages, gas flaring and other externalities of oil exploration pose to their livelihood and survival, the youth have given these companies and the government ultimatums and engaged security forces in combat. They have held public protests and rallies that have attracted international attention. In 1990, Ken Saro-Wiwa, then leader of MOSOP, successfully mobilized tens of thousands of Ogonis to protest against the policies of the federal government in relations to oil wealth and the activities of SPDC, the oil multinational that produces about half of Nigeria's oil. The persistent protests at its facilities forced it to close operations in Ogoni land.

Youths in oil-producing communities have engaged government security forces in combat and have forcibly seized oil installations, production facilities and flow stations. Their unrest has become a regular feature of the oil-producing communities of the Niger Delta. This has taken the form of sabotage (allegedly causing spills and then negotiating what should be paid as compensation for those spills), hostage-taking and harassment (of staff of oil companies), and general disruption of oil production. These violent forms of protest are usually localized. However, they enjoy local and international media coverage because of their impact on oil production. These activities have caused the temporary or permanent cessation of operations in some locations.

During military rule, these communities had to contend with state governors who, being on military posting, were used as agents of the national government to oppress and suppress opposition in any form. This has, however, changed since the return to democratic rule. Under pressure from oil-producing communities, the governments of oil-producing states have begun to challenge the national government's exclusive control of natural resources. But again they have had to deal with the disadvantage of their minority status in the democratic contest of numbers.

Like the social movement, the Summit of Governors and Members of the National Assembly from the South-South Geo-Political Zone have pressed for state control of resources. They want the Land Use Act and other obnoxious laws, which empower the federal government to control the natural resources found in the territories of their communities, abolished. They also contest the distinction between offshore and onshore oil in the implementation of the 13 per cent derivation revenue allocation to oil-producing states by insisting that offshore oil belongs to the communities. (The federal government maintains that offshore resources belong to the federation.) They have pursued this position through a series of public declarations and communiqués. A Bill was tabled before the House of Representative on 9 May 2001 by Senator Harriman of Delta State and 13 others requesting the amendment of the Petroleum Act to compel oil companies to site their headquarters in their main areas of operation, and vest the ownership and control of petroleum resources in the oil-producing states, local governments and communities, thus reversing the spirit of the extant laws, among others to reduce tension, poverty and violence in the oil producing communities. The Bill caused a tempestuous session and was thrown out with 81 against 64 votes in favour, along a sharp north–south divide (see Aiyede 2005).

Another development of note is the federal government's petition, asking the Supreme Court to declare that 'the natural resources located within the exclusive economic zones and the continental shelf of Nigeria are, subject to the provision of any treaty or other written agreement between Nigeria and any neighbouring littoral foreign state, derived from the federation and not from any state' (Djebah 2001: 9) – clearly, a response to the demand by oil-producing states in the country that there should be no dichotomy between offshore and on-shore oil in the calculation of the 13 percent revenue allocation based on derivation.

In a decision made on 5 April 2002, the Supreme Court declared that the littoral states could not legally seek to control natural resources located beyond their seaward boundary. It also declared unconstitutional however the federal government's refusal to begin sharing the 13 per cent derivation formula from May 1999, and First Line Deduction System (FLDS), the latter a procedure whereby the federal government first deducts a percentage of funds credited to the federation account for the payment of debt before sharing the balance among the federal, state and local governments, among others. The first decision threatened to aggravate the conflicts in the oil-producing communities such that President Obasanjo had to adopt a political solution by presenting an appropriate Bill to the national assembly. This Bill, which was signed into law in early 2004, abolished the distinction often made between offshore and on-shore oil in the distribution of revenue. Meanwhile, the 1999 Constitution on which the Supreme Court based its decisions is undergoing a review. Many politicians have expressed the opinion that the issue should, as a result, be treated as a political rather than a legal/constitutional matter before the federal government instituted the suit.

Global civil society and local struggles

The global movement for environmental justice and sustainable human development has not only provided opportunity for drawing global attention to the untoward activities of transnational oil companies in Nigeria, it has added voice to the concerns of these oil-producing communities and their organizations. Also, it has inspired the formation of environmental-focused NGOs in Nigeria. Several human rights organizations have been influenced, such that they began to emphasize environmental issues and the deleterious activities of the oil corporations. The result of these developments has been the publicizing of social and ecological risks and dangers posed by the way oil exploitation is carried out in Nigeria. In the event, the role of the Nigeria state and its collusion with transnational oil corporations in perpetrating environmental degradation, repression of resisting communities and disregard for the livelihood of the peoples of these regions has become publicly known. Indeed, NGOs have been the major agencies for making public the degree of degradation, irresponsible behaviour of oil companies and the violence that has characterized oil exploration in the Niger Delta.

These organizations have utilized their links with the media, their direct and indirect research capacity to unveil the realities in the Niger Delta. Thanks to their outputs, scientific data on the overall and long-term effects of environmental degradation on the livelihoods of those who live in the region are just beginning to emerge (Ojo 2002). The most authoritative work is that done by the Human Rights Watch (1999), *The Price of Oil: Corporate Responsibility and Human Rights Violations in Nigeria's Oil-Producing Communities*. This has been supported by a series of publications in human rights assessments by local rights-based NGOs. For instance, *Boiling Point* by the Committee for Defence of Human Rights (CDHR) published in 2000, Ken Saro-Wiwa's *The Crisis of the Nigerian State* published in 1998, and the series of Annual Reports of the Civil Liberties Organizations. The Internet has also been utilized to publicize the reports.

Many of the local publications were the outcomes of conferences and workshops organized to debate the issues and map out strategies for promoting social justice in the region, thus putting pressure on both the Nigerian state and the oil corporations.

These organizations have been involved in monitoring oil spills, sensitization of the peoples and communities on environmental rights, publishing periodicals to disseminate information on environmental issues and generally organizing protest rallies against human rights abuses and practices that contravene the ethics of sustainable development. They have instituted court cases on behalf of victims and have written letters of appeal to government, oil companies and communities as the case warrants in the interest of equity, justice and peace (see Ikporukpo 2000).

These organizations have utilized the opportunity offered by the internationalization of environmental interest groups, such as access to environmental rights and Africanist groups in the north to mobilize public opinion in major Western cities against multinational oil corporations. In the United States and Britain, for instance, a strong coalition of NGOs coalesced to challenge the Abacha government and oil companies. Activists picketed Shell petrol stations and carried out mass civil disobedience and direct action protest at Shell facilities in major cities. These worldwide responses had a substantial impact on the oil companies which have began to show increasing regard for human rights principles, at least in their rhetoric (Fleshman 2000: 186–7).

Many of these protests enjoyed widespread media coverage. Indeed, Fleshman refers to the dramatic impact of the coverage of Chevron activities by Pacifica Radio Reporter Amy Goodman, involving the use of Chevron helicopters to fly mobile policeman to its platform in 1998 to remove unarmed youths. Two young men were shot dead in the incident (ibid. 189–90).

Conclusion

The change in the global human rights context and the opportunity of globalizing civic values and global action in its defence provide real hope for putting transnational risk on the global governance agenda. For less developed contexts, state weakness and financial pressures often lead to the disregard of local communities and local environmental movements. This in turn forces local communities and organizations to adopt costly violent behaviour in protest against environment degradation and other social injustice that straddle natural resource exploitation. Thus, the situation of risk is further worsened and the management made more complex. As we see from the analysis of the basis and forms of interaction by actors in oil exploitation in Nigeria, there is a clear challenge at the global level for risk analysis and management.

There is a critical role for global civil society because of the relative weakness of communities and peoples in relation to global corporations and national states whose interests coincide in the process of accumulation through natural resources exploitation. Global civil society holds the potential for promoting the realization of social justice for oppressed local minority communities in less developed countries, and thereby help improve transnational risk management. This can be realised through increased capacity to impact on global governance institutions and to understand and represent local problems and peoples at the global level. These carry both activist and institutional challenges requiring a rethinking of the direction of global governance in terms of transnational risk management.

References

Adejumobi, Said and Adewale Aderemi (2002) Oil and Political Economy of the Nigerian Civil War and Its Aftermath, in Egosa E. Osaghae, Ebere Onwudiwe and Rotimi Suberu (eds.) *The Nigerian Civil War and its Aftermath, Ibadan: Programme on Ethnic and Federalism Studies*, pp. 191–206.

Aikhomu, A. A. (1996) Federal–State Relations under Military Government, 1985–1992, in J. I. Elaigwu, P. C. Logams and H. S. Galadima (eds.) *Federalism and Nation Building in Nigeria: The Challenges of the 21st Century* (Abuja: NCIR, 1996), p. 51.

Aiyede, E. (2002) *Decentralising Public Sector Collective Bargaining and the Contradictions of Federal Practice in Nigeria*, African Study Monograph, (Kyoto). 1 (2): 11–29.

Aiyede, E. R. (2005) Intergovernmental Relations and the Strengthening of the Nigerian Federation, in Ebere Nwudiwe and Rotimi Suberu (eds.) *Nigerian Federalism in Crisis: Critical Perspectives and Political Options. Ibadan: Programme on Federalism and Ethnic Studies*, pp. 220–230.

Ashton-Jones, N. (1998) *The Ecosystems of the Niger Delta: An Environment Rights Actions Handbook* (Ibadan: Kraft Books).

Djebah, Omah (2001) Resources Control: South–South Resolves to Tackle Federal Government on All Fronts. *The Guardian*, 26 February.

Douglas, O. (1998) *Shaming Eco-Destroyers in Environmental Action in Nigeria's Niger Delta*, October/December.

Edwards, Michael, David Hulme and Tina Wallace (1999) *NGOs in a Global Future: Marrying Local Delivery to Worldwide Leverage*. Background paper (mimeo) to NGOs in a Global Future Conference, 10 January, University of Birmingham.

Fleshman, Mike (2000) The International Community and Crisis in the Oil Producing Countries, in Wumi Raji, Ayodele Ale and Eni Akinsola (eds.) *Boiling Point* (Lagos: Committee for the Defence of Human Rights), pp.179–195.

Fone, Martin and Peter Young (2000) *Public Sector Risk Management* (Butterworth: Heinemann).

Gowon, Yakubu (1996) Federalism and Nigerian Unity: Problems and Prospects, in J. I. Elaigwu, P. C. Logams and H. S. Galadima (eds.) *Federalism and Nation Building in Nigeria: The Challenges of the 21st Century* (Abuja: NCIR), pp. 23–28.

Graf, W. (1988) *The Nigerian State, Political Economy, Class and Political System in the Post-Colonial Era* (London: Heinemann).

Ikporukpo, C. O. (2000) *Environmental Movements in the Niger Delta Region and the Nigerian State*. Centre for Research and Documentation, Kano/Ford Foundation Research project report.

Ikporukpo, C. O. (2002) In the Name of Oil: The Nigerain State and the Niger Delta Crisis', pp. 207–215 in Eghosa E. Osagae, Ebere Onwudiwe and Rotimi Suberu (eds.) *The Nigerian Civil War and its Aftermath*. Ibadan: Programme on Federalism and Ethnic Studies.

Iyayi, Festus (2000) Oil Corporations and the Politics of Relations in Oil Producing Countries, in Wumi. Raji, Ayodele Ale and Eni Akinsola (eds.) *Boiling Point* (Lagos: Committee for the Defence of Human Rights), pp. 151–178.
Joseph, R. (1987) *Democracy and Prebendal Politics in Nigeria* (Cambridge: Cambridge University Press).
Khan, Sarah Ahmad (1994) *Nigeria: the Political Economy of Oil* (Oxford: Oxford University Press).
Lewis, P. (1997) From Prebendalism to Predation: the Political Economy of Decline in Nigeria, *Journal of Modern African Studies*, 37: 79–103.
Obi, C. (1995) Oil Minority Rights versus the Nigerian State: Environmental Conflict, its Implications and Transcendence. Paper presented to CODRESIA 8th General Assembly and Conference, Dakar, Senegal, 26 June–2 July.
Obi, C. (1997) *Oil, Environmental Conflict and National Security in Nigeria: Ramification of the Ecology-Security Nexus for Sub-Regional Peace*. Occasional Paper, Arms Control, Disarmament, and International Security Program, University of Illinois of Urbana-Champaign.
Obi, C. (1998) The Impact of Oil in Nigeria's Revenues Allocation System: Problems and Prospects, in Kule Amuwo, Adigun Agbaje, Rotimi Suberu and George Herarnlt, *Federalism and Political Restructuring in Nigeria* (Ibadan: Spectrum Books), pp. 261–275.
Ojo, O. J. B. (2002) *The Niger Delta: Managing Resources and Conflicts, Development*. Policy Centre Research Report No. 49 (Ibadan: Development Policy Centre).
Okonta, I. and O. Douglas (2001) *Where Vultures Feast, 40 Years of Shell in the Niger Delta* (New York: Sierra Club in Conjunction with Random House).
Osaghae, E. (1995) The Ogoni Uprising, Oil Politics, Minority Agitation and the Future of the Nigerian State, *African Affairs*, 94: 325–344.
Pearce, Jennyu (2000) *Development, NGOs, and Civil Society: The Debate and its Future*, Debora Eade ed. *Development NGOs and Civil Society* (Oxfam: Development in Practice Book Series).
Risse-Kappen, Thomas (1995) Bringing Transnational Relations Back in: Introduction, in T. Risse-Kappen (ed.) *Bringing Relations Back in: Non-State Actors, Domestic Structures and International; Institutions* (Cambridge: Cambridge University Press).
Suberu, Rotimi T. (1994) 1991 State and Local Government Reorganisations in Nigeria, *Travaux et Documents*, No. 41.
United Nations Development Programme (UNDP) (1999) *UNDP: Human Development Report 1999* (New York: UNDP).

Part III
Public Health: Limits of the Precautionary Principle

Part III

9
The Responsibilities of the Media in Dealing with HIV/AIDS in Southern Africa

Brett Davidson

Introduction

As the UNAIDS 2004 Report on the Global AIDS Epidemic points out, AIDS is one of the major crises facing the globe, and is 'both an emergency *and* a long-term development issue (UNAIDS 2004a: 3; emphasis in original). It is arguably the most important health and security risk facing Southern Africa. UNAIDS reports that an estimated 25 million people are living with HIV in sub-Saharan Africa, compared with 7.4 million in Asia, 1.6 million in Latin America, 1.3 million in Eastern Europe and Central Asia, and 1.6 million in all the high-income countries combined. The epidemic is worst in Southern Africa, where the prevalence rates are: Botswana, 37.3 per cent; Lesotho, 28.9 per cent; Mozambique, 12.2 per cent; Namibia, 21.3 per cent; South Africa, 21.5 per cent; Swaziland, 38.8 per cent and Zimbabwe, 24.6 per cent (UNAIDS 2004a). HIV/AIDS is far more than simply an issue of domestic public health for Southern African countries. It has been named the 'most fundamental cause of the Southern African crisis' (Farlam 2003) and an 'emergency development challenge' for the entire region (Chirambo and Caesar 2003: 5). Despite this the media, specifically in South Africa, have persisted in focusing on HIV/AIDS primarily as a health matter, and almost exclusively in national or domestic terms, with the result that many crucial dimensions of the epidemic have received little attention and emphasis within the public sphere.

This chapter examines the nature of HIV/AIDS as a transnational risk, and outlines some of the epidemic's impacts on democracy and human security in Southern Africa. It further explores the role of the media in covering HIV/AIDS, and an assessment is made of the South African media's performance in this regard, on the basis of existing studies as well as the author's personal correspondence and conversations with five experts and role players working in the fields of media, civil society and HIV/AIDS.[1] Finally, some solutions for poor media performance are explored, in the context of constraints and unresolved debates which must be taken into account in seeking a way forward.

HIV/AIDS as a transnational risk in Southern Africa

While HIV prevalence rates continue to rise, millions of people infected years ago are now falling ill and dying, and all countries in the region are experiencing a full-scale AIDS epidemic. HIV infection has been referred to as the first wave of the epidemic, with the subsequent illnesses and deaths from AIDS seen as the second wave. As a consequence, a third wave is only just beginning to be felt. This third wave is the social, economic and political impact of HIV and AIDS (De Waal, cited in Chirambo and Caesar 2003: 7). It is worth taking a closer look at some of the multiple dimensions of this so-called third wave, and the risks it carries.

First, the HIV/AIDS epidemic presents threats to democracy. According to Mattes (2003) there are three factors that are required to ensure sustained democracies:

1. a healthy political economy, economic growth and low or falling levels of economic inequality;[2]
2. strong political institutions; and
3. a supportive political culture in which citizens believe in democracy and actively participate in public life.

HIV/AIDS is placing all three factors under threat in Southern Africa. For example, the World Bank has warned that the South African economy faces collapse within three generations unless the government changes its approach to the pandemic (Bell, Devarajan and Gersbach 2003). AIDS is killing large numbers of agricultural workers in Southern Africa, thus crippling the agricultural sector – a mainstay of the economy in most countries. In addition to this, because HIV/AIDS is affecting poorer people and households disproportionately, it is leading to a further widening of the gap between rich and poor. For example, studies in South Africa and Zambia have shown that monthly incomes in poor households affected by AIDS fell by 66–80 per cent because of the burden imposed by having to cope with AIDS-related illnesses (UNAIDS 2004a: 9).

While economies are under threat, political institutions and social services face attrition as skilled and experienced personnel succumb to the disease. A study by Manning (2003) of the impact of HIV/AIDS on the eThekwini municipality in Durban, South Africa, shows that the epidemic has already begun reducing the capacity of many municipal departments to deliver on their mandates. At the same time, it is beginning to change patterns and levels of demand for municipal services in ways that are as yet unpredictable and thus difficult to plan for. Malawi is struggling to hold increasing numbers of expensive local by-elections as councillors die from AIDS-related illnesses (Ngwembe 2003). In countries where improvements in education are a crucial part of development plans, AIDS is contributing to shortages of teachers, as educators become ill and die (UNAIDS 2004a). Similar effects are being felt in the health and other critical social sectors.

As for Mattes' (2003) third requirement of a political culture supportive of democracy, the impact of HIV/AIDS on public participation in democracy and governance is uncertain.[3] As institutions falter in delivering on their mandates, however, people may begin to lose faith in democracy and elected government. A further factor is that in many cases HIV/AIDS is increasing governments' dependence on foreign aid, thus placing them under increased pressure to meet the demands and conditions of the superpowers and multilateral institutions. This means a corresponding reduction in the power of African citizens to control their own governments' political and economic decision-making (Chirambo and Caesar 2003: 19). If citizens believe that their preferences and choices exert little influence on government policies and actions, they are likely to become cynical and apathetic about democracy.

In addition to its threats to democracy, HIV/AIDS has an impact on security. Contemporary discourse has moved from a focus on state security to one on the security of people. In this context, 'human security' refers to the protection of vital freedoms, safety from hunger, disease and repression, and protection from sudden threats to the daily life in the home, at work and in communities (Chirambo and Caesar 2003; Ogata and Cels 2003). Apart from its threat to the health of individuals, HIV/AIDS is disrupting families and entire communities. The epidemic is presently posing a serious threat to food security by compounding the impact of drought and agricultural mismanagement in the region. An increase in the number of orphans and a rise in child-headed households mean that large numbers of children are growing up without effective parental guidance and control (Schönteich 2003). This can lead to social breakdown and an increase in crime. More broadly, as a contentious public issue, HIV/AIDS may intensify political polarization and provoke social fragmentation. Furthermore, as state institutions weaken as the result of HIV/AIDS – including institutions directly concerned with security, such as the armed forces and police services – opposition groups in very divided societies could try to exploit these weaknesses, even going as far as to instigate civil unrest or stage coups. This danger is very real in a number of Southern African countries (Chirambo and Caesar 2003: 22).

Third, HIV/AIDS has a gender-related impact. The path of the epidemic is influenced by gender inequalities, while it also impacts on, and threatens to exacerbate further, gender inequalities and skewed power relations. Chirambo and Caesar argue that HIV/AIDS has become 'a strong influence in the construction of masculinity and male sexual behaviour' (2003: 26) and that as a result, many men are experiencing a crisis of identity. But the primary impact is on women. For a range of reasons – physiological, social, cultural and economic – women are more at risk than men from HIV infection and an early onset of AIDS (Chirambo and Caesar 2003). For example, in the home, women who attempt to initiate safer sexual practices often fall victim to domestic violence, while female refugees are many times more likely to be infected with HIV than members of the general population.

In turn, women and girls are also particularly vulnerable to the negative social consequences of HIV/AIDS:

> Girls drop out of school to care for sick parents or for younger siblings. Older women often take on the burden of caring for ailing adult children ... Older women caring for orphans and sick children may be isolated socially because of AIDS-related stigma and discrimination. Stigma also means that family support is not a certainty when women become HIV-positive; they are too often rejected, and may have their property seized when their husband dies. (UNAIDS 2004a: 8)

HIV/AIDS as a transnational risk

According to Bettcher and Lee (2002), the concept 'transnational' is used to mean activities (or risks) that cross state as well as national boundaries. In other words, in addition to activities that literally span international borders, the concept is widely used to refer to activities such as the use of the Internet, or financial flows across states – activities that in fact render borders meaningless and are not easily controlled by governments, if at all.

The HIV/AIDS pandemic in Southern Africa is just such a transnational risk. Numerous regional factors and cross-border activities have an impact on the spread of the pandemic. One such transnational factor is cross-border migration, which involves large numbers of people and takes many forms. This includes decades-old patterns of migrant labour and trade-related mobility, both within and between countries in the region. An additional factor contributing to the spread of HIV/AIDS and complicating efforts to combat it is a significant degree of politically-related conflict and violence, which in turn contributes to migration (Chirambo and Caesar 2003; Schönteich 2003).

Responses to the pandemic, formulated by governments and civil society at a national level, are influenced and play out in the context of globalization, international agreements and a rapidly developing and changing international communications environment. Birdsall (2003) argues that in the context of globalization, risks such as communicable diseases entail disproportionately high costs for poor countries and poor people. In some countries, the economic requirements imposed by the International Monetary Fund and World Bank have severely limited governments' policy options, often compelling drastic cuts in health budgets, for example. Furthermore, the effects of international agreements on the protection of property rights, as well as the actions and policies of multinational pharmaceutical companies, have a profound influence on the affordability of anti-retrovirals and other drugs. A further factor hampering an effective response to HIV/AIDS is the brain drain of skilled medical professionals from the Southern African region. For example, of the more than 1,600 doctors that Zambia has trained at state expense, fewer than 400 remain in the country (Guy 2003).

Transnational communication technologies such as email and the Internet have also played an important role, with both positive and negative outcomes. This is probably most notable in South Africa, where efforts to educate the public and combat HIV/AIDS have been complicated by remarks made by the President and Health Minister, questioning the link between HIV and AIDS, and casting doubt on the efficacy of anti-AIDS drugs (Grobbelaar 2003). President Thabo Mbeki is believed to have developed his controversial views during late-night sessions spent browsing the Internet, where he encountered the arguments of so-called 'AIDS dissidents'. On the other hand, activist organizations such as the Treatment Action Campaign (TAC) have made very effective use of email and the Internet to mobilize international support and raise funds for their fight to ensure widespread access to anti-retroviral drugs (Jacobs 2003a).

Ogata and Cels (2003) point out that in combating transnational threats to human security, it is crucial that countries work together to combat the pandemic and, in doing so, adopt strategies that link political, humanitarian, development, public health and other aspects in an integrated way. Yet despite acknowledging that HIV/AIDS represents an emergency for the entire region, the member states of the Southern African Development Community (SADC) have yet to develop and implement common policies to deal with the disease (Chirambo and Caesar 2003; van Schalkwyk 2003). Perhaps, despite their public pronouncements, many governments, civil society organizations and other role players have not yet fully grasped the nature of the HIV/AIDS epidemic as the predominant transnational risk facing the countries of the region (De Cock, Mbori-Ngacha and Marum 2002; Whiteside 2002). There are a number of reasons why this might be so. First, the nature of the epidemic, with a long incubation period between HIV infection and the onset of AIDS, has meant that the impact is delayed or remains invisible for extended periods of time. The impacts can also be hidden if AIDS deaths are not recorded as such in official statistics, but attributed to various opportunistic infections.[4] Second, the response to HIV/AIDS and the framework within which it is understood has been influenced to a significant degree by the models and approaches adopted in developed countries such as the US and Europe. Because there are major differences in the nature of the respective epidemics in the West and in developing regions such as Southern Africa, however, Western-influenced frameworks are not necessarily appropriate (De Cock et al. 2002). Third, South Africa, the powerhouse in the region, has failed to take the lead in the fight against HIV/AIDS. Again, this can be attributed to the 'denialist' stance adopted by Mbeki and some members of his Cabinet. This has meant that energy and attention is often sidetracked into arguments about whether HIV does indeed cause AIDS, instead of being focused on finding the best ways to combat the epidemic. Finally, the media in the region have often failed to give the pandemic the attention and prominence it warrants, or have often focused on issues and stories that have contributed to rather unproductive public discourse on the disease (Grundlingh 2001; Ngome 2003).

Media and HIV/AIDS as a transnational risk

There has been widespread attention to the role of the media in combating HIV/AIDS. Much of this attention has focused on the role of the media in educating people about the disease and how to avoid infection. For example, this is the primary emphasis of the UN's Global Media AIDS Initiative. There is an argument that the media's role should be seen as much wider than this, however – that in the context of HIV/AIDS as a multifaceted transnational risk, the media have a key role to play in ensuring that all the various aspects of the epidemic, including policy-related issues, receive attention in the public sphere. This is true even in the realm of prevention, since an effective prevention campaign involves far more than simply educating individuals about safe sex because social and often other conditions make it difficult or impossible for individuals to avoid exposing themselves to the virus. For example, many women are unable to insist on safer sex with their partners because of the threat of domestic violence or because they are economically dependent on men (UNAIDS 2004a). So governments need to undertake social interventions to change attitudes, provide women with better economic opportunities and with means of protecting themselves against violence. Such issues are matters of public policy and social mores, which need to be discussed and deliberated in the public sphere.

In this context, several roles for the media can be identified. First, the media play a role in providing the public with the information they need to make intelligent choices as citizens (Wasserman 2003). This would involve reporting on what elected representatives and public servants are and are not doing to combat the pandemic and mitigate its effects. Citizens also need the media to provide information on the nature of the threat posed by HIV/AIDS and on governments' responses in order to decide whether, and in what fashion, to engage in other activities designed to influence policymakers (including governments and intergovernmental organizations). The media play a role in informing other key role players too. According to Shepperson (2000), for example, health workers use newspapers as an important source of information on HIV/AIDS research.

The media also have an agenda-setting role (Severin and Tankard 1992). Thus, in the way that they tackle the issue, such as the prominence and frequency of stories, the media can influence the salience of HIV/AIDS in the public agenda. Beyond this, however, the media help determine the way in which HIV/AIDS is framed or defined in the public realm (de Cock et al. 2002). As de Cock et al. point out, 'How an issue is defined strongly affects how it is addressed' (2002: 3). This could include, among others, defining the disease as a matter of individual sexual health, merely one of many issues involved in the political power-plays between parties, or as a major transnational risk that requires concerted regional and international co-operation in response. Furthermore, the media has a role as a 'watchdog': to monitor the actions and decisions of governments, public and private organizations, and regional bodies. This 'watchdog' role is important, given assessments that some of the failures of regional bodies in dealing with HIV/AIDS are due in part to a lack of accountability and transparency (Chirambo and Caesar 2003). Finally, the media act as a key constituent of the public sphere, enabling members of the public to exchange ideas, discuss the issues of the day and have their views reflected back to decision-makers.

It is difficult to gain a comprehensive picture of the degree to which the media in Southern Africa have been performing these roles, since there is a dearth of media research in the region. There have been several relatively recent studies of the South African media's coverage of HIV/AIDS however, and an overview may be useful in providing insight into media performance under what can arguably be described as the most favourable conditions in the region.

The South African Constitution and legal framework protect freedom of expression, and the country enjoys a wide array of media products and formats. While these advantages make South Africa the envy of many of its neighbours, it is not to say that conditions are ideal. As Stein (2002) points out, there are several constraints under which the news media operate, and the findings of studies on media content should be seen in the context of these. One of the key trends since 1994 has been an increased concentration of media ownership in the hands of a few companies, along with an increase in foreign ownership.[5] An important result of this is that newsrooms have faced cuts in budgets and staff as companies seek to maximize profits (Foster 2004). This has meant that journalists have become increasingly overstretched and unable to specialize. It has also meant that editors are reluctant to grant reporters time off to attend specialized short courses. Poor salaries and working conditions mean that experienced reporters have tended to move into management or leave journalism for other fields. This has led a shortage of skills and what has been labelled the increasing 'juniorization' of the newsroom (Foster 2004; Felix 2005). Most reporters are under the age of 25 (Stein 2002) and lack the skills and experience to deal adequately with complex and multifaceted issues such as HIV/AIDS. Furthermore, because advertisers have tended to favour media catering to the more affluent (and generally white) audiences, most newspapers and private broadcast media emphasized the issues of concern to these audiences, often neglecting the views, perspectives and priorities of the poor (and mostly black) majority.

According to a study by Grundlingh (2001), early media reports as well as government responses to HIV/AIDS were highly influenced by perceptions gained from the nature of the epidemic in Western countries, where AIDS was seen as a disease primarily affecting homosexual men and intravenous drug users. These perceptions and models blinded the authorities and the public to the real nature of the HIV/AIDS epidemic in Africa, contributed to stigmatization of people with HIV/AIDS and blocked effective responses (Grundlingh 2001). While HIV/AIDS in South Africa is no longer reported in this manner, de Cock et al. (2002) point out that much of the discourse around the epidemic and many of the policy responses are still influenced by that initial framing and prevent policymakers from tackling HIV/AIDS with the necessary intensity and vigour.

Shepperson (2000) analysed HIV/AIDS reporting in all the major South African newspapers over two periods in 1999 and 2000 and concluded that the print media were not doing full justice to the impact and scale of the HIV/AIDS epidemic (2000: 8). He noted that newspapers tended to be reactive rather than proactive in their reporting, with reports generated by organizations and individuals outside the newsroom far outnumbering those based on papers' own inquiry. Government events and spokespeople dominated as the primary sources of stories, and reporters rarely checked facts and statistics provided by official sources. In addition, coverage was dominated by ongoing political controversy over provision of anti-retroviral drugs and the president's questioning of the link between HIV and AIDS and the safety of the drug AZT.

A study by Stein (2002), which entailed qualitative interviews with news media gatekeepers, practitioners and stakeholders, also highlighted the media's emphasis on the controversy over Mbeki's views. While Stein (2002) argued that this issue was of vital national importance, the problem was that media coverage tended to sensationalize the debate in ways that polarized the HIV/AIDS discourse along racial and sometimes party political lines. Aside from the anti-retroviral controversy, Stein (2002) highlights other problems. While the media have moved on from featuring AIDS as a 'gay' epidemic to depicting it as a disease mostly affecting poor black women, coverage often plays into existing racial and gender stereotypes. This can perhaps be partly explained by the fact that the voices and perspectives of those 'infected and affected', particularly the voices of women, are largely absent from media coverage (de Wet 2003). Furthermore, many newspaper editors and journalists consider HIV/AIDS to have low news value for their readers, who are predominantly white and/or from middle and upper income groups (Stein 2002).

When it comes to coverage of HIV/AIDS in a transnational context, Wasserman (2003), Chirambo (2003) and Jacobs (2003b) all feel that the

South African media rarely report on regional initiatives such as Southern African Development Community (SADC) frameworks for combating HIV/AIDS. Jacobs (2003b) states that coverage of HIV/AIDS 'is not integrated as a Southern African story'. Wasserman (2003) says that when South African media do cover HIV/AIDS in the region, they tend to do so in terms of local political conflicts rather than taking on an integrated, regional focus. Chirambo (2003b) argues that even when the media deal with political controversy over the disease, HIV/AIDS is largely reported as a health issue[6] and the interrelationship of HIV/AIDS with broader social issues, such as those of democracy, governance, the economy and human security, are rarely reported. Likewise, there is little debate over appropriate national and regional policy responses to these issues.

There is praise for some aspects of HIV/AIDS coverage in the South African media, however. Several role-players and scholars (Stein 2002; Jacobs 2003b; Valentine 2003; Wasserman 2003) believe that coverage of HIV/AIDS has improved significantly in recent years. While Stein (2002) criticizes the politicization and sensational nature of coverage of the ongoing disputes over treatment, she points out that this may have had unintended positive consequences in compelling citizens to grapple with the complexity of the issues of HIV/AIDS prevention and treatment.[7] It is also in the context of disputes over treatment that certain transnational factors linked to HIV/AIDS are raised, such the actions of multinational pharmaceutical companies and international agreements on intellectual property rights, which have contributed to making vital drugs inaccessible to the poor (Jacobs 2003a).

It is worth bearing in mind that such mixed assessments with respect to coverage of HIV/AIDS are not directed at the South African media alone. For example, in a major study released by the Kaiser Family Foundation in 2004, the American media received a similarly mixed assessment of their coverage of HIV/AIDS between 1981 and 2002 (Brodie et al. 2004). The study found a steady decline in overall coverage of HIV/AIDS in the news over the years, and the authors suggest that this might be the cause of a parallel decline in the American public's perception of the urgency of problem. In addition, other problems were highlighted, such as the fact that specific populations bearing most of the brunt of the epidemic in the US received relatively little media attention. For example, gay men were the focus of 4 per cent of stories overall, minorities the focus of 3 per cent and women the focus of 2 per cent of stories.

Perhaps most interesting in the context of this discussion, however, is the finding that the nature of coverage of HIV/AIDS in the American media changed over the years, with a shift away from portrayal of the epidemic as a domestic health problem, to coverage of HIV/AIDS as a global issue. At the same time, the media have tended increasingly to treat the epidemic as a business and political story. The authors argue that these trends can be seen as encouraging, as they accurately reflect changes in the nature of the epidemic in the US (where AIDS has largely become a manageable chronic disease) and the rise of AIDS as global crisis. At the same time, however, there has been a corresponding decline in the number of stories with an educational element. This may be cause for alarm, since large numbers of Americans still lack knowledge about how HIV is transmitted, and since the numbers of new AIDS cases in the US have recently begun to increase (albeit slightly), after years of being stable.

This study holds important lessons for discussions and assessments of the South African media. Overall, scholars and commentators seem to agree that there is plenty of room for improvement in the media's treatment of the HIV/AIDS epidemic. Despite some positive trends in coverage in recent years, most Southern African citizens and governments 'still don't understand the absolute hideousness of what's happening' (Stein 2003: 17). Stein (2003) and Chirambo (2003) argue that it is vitally important that the media begin contributing to a better understanding of HIV/AIDS in the social, cultural, economic and historical context of the entire region. Wasserman (2003) feels that coverage of HIV/AIDS within a transnational perspective might 'provide for a more nuanced and complex treatment of the subject', while helping individual countries to benefit from insights and solutions arrived at elsewhere. Chirambo (2003) feels that reporting within a transnational context would enable governments' performance to be compared with norms and standards agreed at inter-governmental level. For example, he feels that the media could and should be monitoring whether governments are meeting their agreed commitments to spend at least 15 per cent of their total budget on health. What Brodie et al.'s (2004) study seems to demonstrate, however, is that it is clearly not desirable that reporting from a transnational perspective should displace other kinds of coverage of HIV/AIDS, and education about prevention, treatment and related issues should remain a priority. This discussion of 'desirable' media coverage, however, raises the difficult question of the extent to which the media have obligations to the public when it comes to their treatment of any issue. In the context of their discussion of the US media's coverage of HIV/AIDS, this is what Brodie et al. call the 'ever-present question of the appropriate role of journalists, especially in the context of a public health epidemic: to what extent do the media have a responsibility to educate the public, as opposed to focusing only on reporting the news?' (2004: 7).

Responsibilities of the media?

Do the news media have a responsibility, even a duty, to prioritize HIV/AIDS as a story – actively to place the issue at the top of their news agendas in the light of its importance as the major threat facing the region? Studies in South Africa have shown that editors are divided over whether HIV/AIDS should compete with other stories from day to day on the basis of 'news value', or whether the media have a responsibility actively to prioritize HIV/AIDS (Stein 2002; Anonymous 2003).

This debate has deep roots, as talk about the media's social responsibilities seems to challenge the liberal Western notion of the news media as a 'fourth estate', detached from political and social institutions. Under such notions, the news media should have no obligations other than to report the news as they see it. Thus, some editors and scholars would argue that the media must be free to make their own news judgements, that HIV/AIDS as a story has to compete equally with the myriad of other issues of the day, and that if the 'news' value of HIV/AIDS declines and the issue begins to disappear from front pages or the news bulletins, then so be it. To talk of getting the news media to fulfil their 'responsibilities' by intentionally placing HIV/AIDS at the top of the news agenda or emphasizing certain aspects of the epidemic is to begin to tamper dangerously with press and media freedom.

Such classical views of press freedom have long been brought into question in the developing world in general, and Africa in particular. In the 1970s, for example, the concept of 'development journalism' was proposed in the context of discussions of the role of the media, facilitated by Unesco. Proponents of this idea argued that in the light of the immense challenges facing Africa, journalists had a responsibility to highlight development processes and play an active role in pressing for change (Kariithi 1994). The concept of development journalism is now widely seen as discredited, since most attempts to implement it in practice ended up by bringing the media under government control. Despite the failure of the experiment, however, a number of scholars believe that the arguments for development journalism did succeed in raising some key problems with the classical, liberal views of the role of the media. One of these is the failure of the classical model to deal with the media's relationship to capital. Rønning argues, for example, that classical liberal views of the media are 'far too preoccupied with the relationship to the state, and too little attention has been paid to the relationship to the market, and indeed to the problems of having media which are too dependent on economic [and other] interests' (1994: 19).

Wasserman (2005) feels that while the existence of free and vigorous news media may be in the public interest, it is important to remember that the media also have their own politico-economic interests. Numerous studies in many countries have shown that 'news judgements' are not value-free, but reflect the economic, cultural, racial and other biases of those making them. Thus, a white male editor of a major South African newspaper can justify the relative absence of HIV/AIDS as a story in his paper by arguing that it is not news for his (mostly white, mostly affluent) readers. He can go on to justify the fact that his newspaper is targeted at an affluent white minority by arguing that the paper must make money, and that this is the audience the advertisers pay to reach.

Wasserman (2003) argues that rather than adopting a purely market-oriented approach, the media must accept a responsibility towards society to educate, provide a forum for debate, hold business and government accountable and also work to empower citizens as citizens – not merely view them as the subjects of social marketing campaigns or as media consumers. In the context of HIV/AIDS in Southern Africa, then, this would mean that the media could legitimately be expected to ensure that the type and amount of coverage should accurately reflect the nature of the epidemic as the major crisis facing the region. As Brodie et al. put it, it may not be unreasonable to ask that journalists covering HIV/AIDS actively 'find new ways to keep their audience engaged in a story that may not meet editorial standards for "news" as clearly as it once did' (2004: 7).

Responsibilities of civil society?

It is with this belief that several members of civil society have sought to influence news coverage of HIV/AIDS, to give what they believe is the appropriate emphasis to the epidemic. Perhaps the largest and most prominent attempt to place HIV/AIDS squarely on the media agenda is the UN's Global Media AIDS Initiative, launched in January 2004 (UNAIDS 2004b). As part of this, the UN Secretary-General Kofi Annan publicly appealed to major media organizations to take a more active role in the fight against HIV/AIDS by, among other things, making the fight a corporate priority, providing consistent news coverage of the epidemic, and integrating HIV/AIDS storylines into shows and films (UNAIDS 2004b). Key UN personnel such as Annan and UNAIDS chief Peter Piot have met executives from the major media corporations, and the UN is following up with specific media efforts in Eastern Europe, Russia, India, China, Indonesia and the United States. Aside from the

UN's campaign, numerous global or transnational efforts are underway, or have been going for some time. One is the One World AIDS Radio network, which enables community and other radio stations around the world to share and exchange AIDS-related programming via the Internet. Among other activities, the network runs an annual World AIDS Day competition in cooperation with MTV, which awards prizes to the best audio and video public service announcement on HIV/AIDS (www.oneworld.net).

In South Africa, civil society organizations have worked to improve media coverage of HIV/AIDS by activities such as training journalists, providing Internet-based support for journalists covering HIV/AIDS, establishing specialist news agencies and developing HIV/AIDS reporting guidelines (Shepperson, 2000; Stein, 2002). For example, CADRE has a website which provides access to HIV/AIDS research, and is helping to develop definitive ethical guidelines for reporting on HIV/AIDS.[8] In 1999 a specialist news agency, Health-e, was established to provide health-related reports free to the print and electronic media (Valentine 2003), and it has since won several South African–US health journalism awards in recognition of its work. In addition, various organizations provide training courses on HIV/AIDS for working journalists.[9] While many of these efforts focus on the health-related aspects of HIV/AIDS, some projects do have a wider focus. For example, the Media, Governance and AIDS project of the Institute for Democracy in South Africa (Idasa), aims to improve media coverage of HIV/AIDS as it relates to democracy and governance in the Southern African region. The project holds workshops for journalists and disseminates ready-to-use news and features to the media in collaboration with the Inter Press Service. Interestingly, the project also aims to work with political communicators by developing handbooks and providing them with current research. The rationale for this is that the media in Southern Africa tend to react to the activities and utterances of political leaders, and that it is these leaders who play the primary role in setting the news agenda. Thus, if there are improvements in the quality and scope of what political leaders and commentators say about AIDS, this will have a ripple effect in improving media reports (Chirambo 2003).

Aside from simply seeking to highlight HIV/AIDS as a prominent issue, however, or improve the quality of reporting on the issue in a general sense, many members of civil society seek to push a specific agenda through the media. While editors and journalists grapple with the moral dilemmas around HIV/AIDS reporting, there are numerous role players attempting to sway them in one direction or another. Just as business, government and other sectors seek to influence the media, to place their issues on the agenda, and provide their specific 'spin', so civil society organisations seek to do the same. This is generally seen as commonplace and acceptable. So, for example, it is quite legitimate for an international medical NGO holding a briefing for health journalists on a pilot treatment programme in order to counter statements by government officials questioning the efficacy of anti-AIDS drugs, or in a treatment advocacy group organising a march to Parliament to draw attention to its demands.

Such advocacy activities raise another level of debate about the obligations of the media in reporting on HIV/AIDS, however. It is one thing to argue that the media have an obligation actively to prioritize HIV/AIDS as a story and cover the epidemic in all its many facets. It is another thing altogether to argue that the media should take a specific stance on certain controversial HIV/AIDS-related issues. Yet this has become an important debate for media scholars and practitioners in South Africa. For example, if President Mbeki questions the causal link between HIV and AIDS, espousing what has come to be known as a 'denialist' view, is it the responsibility of the media simply to report the his views, or to point out that his stance has been widely condemned by the orthodox scientific community? What weight should reports give to the dissident views held by the president and other prominent figures? If the South African government delays programmes to provide anti-retroviral drugs to HIV-positive mothers giving birth at state institutions and a civil society organization successfully challenges this in the Constitutional Court (Rickard 2002), should the media take sides, or simply report the views of the two parties? Should the media have campaigned for widespread anti-retroviral treatment, as some media organizations once campaigned against apartheid? Questions like these are constantly being considered and debated in media and academic circles, and there are no simple answers. As Krüger (2005) points out, the AIDS epidemic is providing profound challenges to journalistic ethics and is confronting journalists with tough moral dilemmas that defy easy resolution.[10]

Notes

1   These are: Herman Wasserman, an expert in media ethics at Stellenbosch University; Kondwani Chirambo, manager of the Governance and AIDS Project at IDASA (Institute for Democracy in South Africa); Sean Jacobs, a prominent South African political commentator; Sue Valentine, editor of the health-e news service; and an editor on the Africa desk at the South African Broadcasting Corporation, who wished to remain anonymous.

2 According to Mattes, historical research has shown that democracy has the best chance of surviving in countries where GDP per capita is above $6,000, where the economy is growing and where income inequality as measured by the Gini coefficient is moving towards, or is below, the mean of 0.35.
3 While participation may decline as people weaken as result of AIDS, the epidemic may also lead to increased mobilization and activism as citizens demand that their governments take action to combat the disease, provide treatment, etc. (Manning 2003).
4 This is still a factor in South Africa, where the official statistics agency, Statistics South Africa, has released a report showing a 57 per cent increase in the country's daily death rate between 1997 and 2002. Stats SA has been reluctant to link the increase in deaths to HIV/AIDS, however, attributing the trend to increases in deaths from tuberculosis, influenza and pneumonia (Schoonakker 2005).
5 This is most notable in the print and online media sector. Restrictions on cross-media ownership have limited the impact on broadcast media to some extent.
6 The political conflicts over HIV/AIDS in South Africa are over health-related issues such as treatment plans, or disputes over whether HIV causes AIDS.
7 However, the remarkable success achieved by regional governments working together to combat malaria, in the relative absence of media coverage (Sharp 2003), prompts speculation about the extent to which the widespread and controversial nature of coverage of government responses to HIV/AIDS has been counterproductive.
8 See www.journ-aids.org.
9 Examples are the Institute for the Advancement of Journalism, the Media Training Centre, and the Medical Research Council.
10 On the dissident issue, Krüger argues that: 'Journalists have to be careful in the way they handle dissident views. While no views should be suppressed, accuracy demands that journalists make it clear these are minority or even fringe views' (2005: 13).

References

Afrobarometer briefing paper no 7 (Cape Town: Afrobarometer, July 2003). Accessed 21 November 2003 (http://www.afrobarometer.org/abbriefing.html).
Anonymous editor at the Africa Desk of the South African Broadcasting Corporation (2003). Personal telephone interview, 21 October.
Bell, Clive, Shantayanan Devarajan and Hans Gersbach (2003) *The Long-Run Economic Costs of AIDS: Theory and an Application to South Africa* (Washington, DC: World Bank). Accessed 20 November 2003: http://www1.worldbank.org/hiv_aids/docs/BeDeGe_BP_total2.pdf.
Bettcher, D and K. Lee (2002) Globalisation and Public Health. *Journal of Epidemiology & Community Health*, 56(1): 8–18.
Birdsall, Nancy (2003) Asymmetric Globalization: Global Markets Require Good Global Politics, *Brookings Review*, Spring, 21(2): 22–27.
Brodie, Mollyann., Elizabeth Hamel, Lee Ann Brady, Jennifer Kates and Drew E. Altman (2004) AIDS at 21: Media Coverage of the HIV Epidemic 1981–2002. Supplement to the March/April issue of *Columbia Journalism Review*.
Carey, James (2002) Personal interview conducted in New York, May.

Chirambo, Kondwani (2003) Personal e-mail to author. Accessed 20 October.

Chirambo, Kondwani and Mary Caesar (2003) *AIDS and Governance in Southern Africa: Emerging Theories and Perspectives* (Pretoria: IDASA).

De Cock, Kevin M., Dorothy Mbori-Ngacha and Elizabeth Marum (2002) Shadow on the Continent: Public Health and HIV/AIDS in Africa in the 21st Century. *Lancet* 360 (9326): 67–74.

De Wet, Gideon (2003) *Agenda Setting Politics: The Voices of the Infected and Affected HIV/AIDS News Sources*. Paper presented to the Media in Africa conference, Stellenbosch, South Africa, 12 September.

Farlam, Peter (2003) AIDS Turns SADC Food Crisis into Major Disaster. *SADC Barometer*, March: 7.

Felix, Bate (2005) News Editors Lack Skills, Sanef Audit Finds. Journalism.co.za 27 May. Accessed 30 May 2005 (http://www.journalism.co.za/).

Foster, Douglas (2004) Letter from Johannesburg: The Trouble with Transformation. *Columbia Journalism Review*, September/October. Accessed 30 May 2005 (http://www.cjr.org/issues/2004/5/foster-letter.asp).

Grobbelaar, Silla (2003) It's Downhill From Here, Unless ... *eAfrica*, August: 10.

Grundlingh, Louis (2001) 'A Critical Historical Analysis of Government Responses to HIV/AIDS in South Africa as Reported in the Media, 1983–1994. Seminar presented at Rand Afrikaans University, Johannesburg, 25 May.

Guy, Trish (2003) Africa's Health Care Crisis. *eAfrica*, September: 14.

Jacobs, Sean (2003a) 'What Do the Poor and Their Organisations Do?' in unpublished PhD dissertation (Birkbeck College, University of London).

Jacobs, Sean (2003b) Personal e-mail to author. Accessed 19 October 2003.

Kariithi, Nixon (1994) The Crisis Facing Development Journalism in Africa, *Media Development* 4: 28–30.

Krüger, Franz (2005) AIDS, News, Views and Journalism, *The Star*, 3 February: 13.

Manning, Ryann (2003) *The Impact of HIV/AIDS on Local-Level Democracy: A Case Study of the eThekwini Municipality, Kwazulu-Natal, South Africa*. CSSR Working Paper No. 35, April.

Mattes, Robert. (2003) *Institutional Mandates and the Challenges of HIV and AIDS: The Potential Impact of AIDS on Democracy in Southern Africa*. Presentation to the Regional Governance and AIDS Forum. Cape Town, 2 April.

Ngwembe, M. (2003) *Southern African Development Community Electoral Commissions' Forum (SADC-ECF): The Role of the SADC-ECF in Mitigating the Impact of HIV and AIDS on Elections and Electoral Processes*. Presentation to the Regional Governance and AIDS Forum. Cape Town, 3 April.

Ngome, Joseph (2003) Reporting on HIV/AIDS in Kenya, *Nieman Reports*, Spring, 57(1): 44–45.

Ogata, Sadako and Johan Cels (2003) Human Security: Protecting and Empowering the People, *Global Governance* 9(3): 273–283.

Rickard, Carmel (2002) State Routed in Constitutional Court's Nevirapine Judgment, *Sunday Times*, 7 July. Accessed 21 February 2005 (http://www.suntimes.co.za/business/legal/2002/07/08/carmel01.asp).

Rønning, Helge (1994) *Media and Democracy: Theories and Principles with Reference to an African Context* (Harare: Sapes Books).

Rønning, Helge (1999) The Unholy Alliance – International Media and the NGOs, *International Public Relations Review*, 21(3): 4–10.

Rosen, J. (2002) Personal interview conducted in New York, 3 May.

Schönteich, M. (2003) *Security, HIV and AIDS: The Impact and Mainstreaming HIV & AIDS within Security Institutions*. Presentation to the Regional Governance and AIDS Forum. Cape Town, 2 April.

Schoonakker, Bonnie (2005) Stats SA Treads Carefully on AIDS, *Sunday Times*, 20 February: 11.

Severin, Werner J. and James W. Tankard (1992) Agenda Setting, in *Communication Theories: Origins, Methods and Uses in the Mass Media* (New York: Longman).

Sharp, Brian (2003) Victories against Malaria, *eAfrica*, May: 11.

Shepperson, Arnold (2000) *HIV/AIDS Reporting in South Africa: an Analysis of the Response of the Press*. Research commissioned by the HIV/AIDS and STD Directorate, Department of Health, South Africa.

Sirianni, Carmen and Lewis Friedland (2001) *Civic Innovation in America* (Berkeley: University of California Press).

South African Council of Churches (2005) *Condoms an Essential Component of Anti-AIDS Strategy, SACC Warns*. Press statement released 4 February.

Stein, Joanne (2002) *What's News: Perspectives on HIV/AIDS in the South African Media* (Johannesburg: Cadre). Accessed 20 November 2003 (http://www.cadre.org.za/pdf/Whats per cent20news.pdf).

UNAIDS (2004a) *Report on the Global AIDS Epidemic* (Geneva: UNAIDS).

UNAIDS (2004b) *Global Media AIDS Initiative* (Geneva: UNAIDS).

Valentine, Sue (2003) Personal interview conducted 23 October in Cape Town, South Africa.

Van Schalkwyk, Gina (2003) More of the Same in Dar es Salaam? *SA Foreign Policy Monitor*. August/September: 1–2.

Wasserman, Herman. (2003) Personal e-mail to author. Accessed 21 October.

Wasserman, Herman (2005) Media is not Only a Mirror, *Mail & Guardian*, 18–24 February, p. 23.

Whiteside, Alan (2002) *Sectoral Impact: What We Know, Don't Know and Need to Know: The True Cost of AIDS*. Presentation to the 14th International AIDS Conference. Barcelona, 7–12 July.

10
A Comparative Analysis of Risk Perception Related to Human Health Issues

Frederic E. Bouder

Introduction

The purpose of this chapter is to explore and compare different levels of acceptance and resistance to risk associated with human health, with respect to the specific cultural contexts of France and the UK. It is developing a preliminary analysis of the perception of controversial vaccines, namely measles, mumps and rubella (MMR) and hepatitis B. The degree of societal acceptance for health-related risks is low compared to other risks. This is especially true of those risks affecting children's health. The number of recent scandals related to health and sanitary issues, such as bovine spongiform encephalopathy (BSE), the blood transfusion scandal and foot and mouth disease, are calling for detailed attention to those factors that may influence acceptance or rejection of risk. Although the level of tolerance of the risks associated with pharmaceutical products seems to have been higher than the acceptance of other health-related hazards, the pharmaceutical sector is at a turning point and vaccines have represented one of the 'pioneer' areas of growing concern. The selection of newspaper articles covering the MMR and hepatitis B controversies, which underpin this chapter, tends to confirm that declining patterns of tolerance for vaccines are a trend and that they are amplified by poor policy responses. Risk mismanagement and crisis avoidance have a critical negative impact on trust and this influences risk acceptance. The case study calls for a dramatic effort towards a renewal of risk communication in the area of risks associated with pharmaceutical products.

It can be argued that Western societies[1] have growing expectations of decision-makers and firms to establish a 'risk-free environment', and hardly tolerate risks unless they consider that they understand them, their probability and their potential effects. For about 50 years academics have compared various attitudes towards risks.[2] On both sides of the Atlantic key areas have been subject to 'precautionary' regulations, reflecting aversion to risk in a significant number of cases (Wiener and Rogers 2002). It appears that we live in a 'post-trust' society where risks have become an intrinsic part of decision-making and regulation processes (Löfstedt 2005). However, the view that risk-aversion would equally affect all human activities would be misleading, as it does not draw sufficient attention to the considerable variations of individual and public perception among distinct risks. In fact, risk research tends to indicate that risk perception is heavily influenced by the nature of the activity concerned and that 'a level of risk that is acceptable for one activity might seem horrendously high or wonderfully low in other contexts' (Fischhoff 1994: 2). And we have known for decades that these differences can be represented (Slovic et al. 1981; 1987). Variables, which affect significantly acceptance and reluctance, include individual and collective factors identified from different theoretical angles, e.g. knowledge theories, personality theories, economic theories or cultural theories (Wildavsky and Drake 1990).

It has been observed that the degree of societal acceptance of health-related risks is generally low compared to other risks. This is especially true of those risks affecting children's health, for example risk resulting from the release of chemicals, and which as a consequence have been strictly regulated in many countries (Löfstedt 2003). The number of recent scandals related to health and sanitary issues constitutes, therefore, a fascinating field of investigation and health-related scandals are calling for detailed attention to those factors that may influence acceptance or rejection of risk.

Looking at risk perception from a trans-boundary perspective also raises a number of specific observations:

- Most of the undertakings which have a critical impact on human health can no longer be understood and analysed in a purely national context. This is equally true for a wide range of activities from agro-food to chemical industry or pharmaceuticals products. These activities are developed and commercialised by global companies for various national markets, they are regulated by a mix of international and national regulations, and their impact is to be observed at a regional and global level.
- In addition to the phenomenon of globalization of potential benefits and risks, one has to raise the issue of whether risk perception is also

being reshuffled. Is it, for example, possible to make generalizations about the perceptions of specific risks? Would these be independent of national contexts? Or are these activities subject to cultural variation?

The purpose of this chapter is to explore and compare different levels of acceptance and reluctance to risk associated with human health, with respect to the specific contexts of France and the UK. We are developing an analysis of the perception of two controversial vaccines, MMR vaccine and hepatitis B. This analysis is based primarily on a media selection from the vast number of newspaper articles covering these scandals. Concrete steps to improve risk communication and prevent future failures are also indicated.

Perceptions of risk related to health in France and the UK: learning from recent events

Health and risk issues in France and the UK cannot be separated from the legacy of recent scandals. France and the UK have been confronted to a number of major crises: the BSE (bovine spongiform encephalopathy) scandal in the UK and France, foot and mouth disease in the UK, the blood transfusions scandal in France, as well as the asbestos scandal in France. These events, in particular the BSE crisis, have been followed by extensive media coverage and passionate political debate, and have resulted in a number of court cases (Powell and Leiss 1997). They were critical in undermining public trust in the regulator, with wider policy consequences (Löfstedt 2005: 16). Striking similarities should be highlighted between these various 'policy failures': France and the UK seem to have been equally unable to anticipate public concern in sensitive areas. Both countries have failed to prevent the release of hazardous substances into the environment, have been unable to enforce acceptable standards of food and health safety, and have not been sufficiently effective in making immediate and adequate responses to pressing health challenges. Observation of these events also suggests that both counties have been subject to a number of 'risk-amplification' phenomena.[3] It should be noted that risk amplification of high-consequence risks and their mediation by expert systems and transmission by the mass media are inseparable from contemporary developments in democracies (Giddens 1991). In the case of the above-cited scandals, poor management and inadequate risk communication have played a critical role (Powell and Leiss 1997).

In addition to risk amplification, these examples tend to show that what we would call a phenomenon of 'risk contamination' took place from country to country as well as from issue to issue. We would suggest defining risk contamination as the translation of a risk from one context to another, especially from one particular field of activity to another, or from country A to country B. The blood transfusion scandal in France and the BSE scandal in the UK appear to have had a key impact on other areas of risk related to human health, and this led to the rapid growth of risk-aversion in both countries, increasing social demands for quick action from decision-makers, as well as declining degrees of trust in science and government. Some academic observers have suggested that we should talk about 'post-trust societies' (Löfstedt 2005). It may well be that 'risk contamination' played a role in further diminishing the levels of trust vested in the regulator. The relevance of risk contamination from one issue to another is supported by the intensity of the controversy following the outbreak of foot and mouth disease in the UK. It could be described as a relatively low-ranking health issue, which turned out to be a major scandal.[4] For 'health experts' and professional risk assessors, foot and mouth disease is a relatively benign pathology. However, societal concerns have been extremely high in the UK, as a result of the 'societal trauma' created by the BSE crisis, which represents 'the biggest failure in UK public policy since the 1956 Suez Crisis'.[5]

The consequences of the BSE crisis were much more than an adjustment to the previous organization of food and safety regulation in the UK and elsewhere in Europe. It had longer-term consequences for the perception of food-related risks. In the UK, it played a role in raising the level of risk-aversion as well as in the development of 'food scares', ranging from salmonella in eggs, listeria in soft cheeses, genetically modified food and contaminated salmon.[6] The BSE crisis has also modified the attitude of all Europeans to food. To some extent, for example, the BSE crisis may have contributed to the stigmatization of GM technology (Vogel 2003).

Although more limited, there are indications that contamination effects of a trans-boundary nature could take place between France and the UK. The BSE scandal in the UK seems to have created high concerns for health and food safety in France, which materialized even after a ban on the importation of British beef was imposed, by a rapid and sustained decrease of beef consumption, an in-depth media coverage of the issue, and an increasing questioning of mainstreamed agricultural production methods. French worries could stem from a series of factors that will not be examined here, but may include a mix of media 'scaremongering', cultural attitudes towards food, perceptions of the levels of safety of French and British products, and high public concern for the sustainability of French beef production both from a safety and economic perspective.

For a long time societal attitude towards pharmaceutical products have followed a quite different path. In the 1950s and 1960s a series of pharmaceutical scandals caused patients to think twice about the drugs they were being prescribed. Diethylstilbestrol (DES) was given to pregnant women at risk of miscarriage and caused genital deformities and cancers in their children. Thalidomide, another medication prescribed to pregnant women, also caused serious deformities. The most tangible consequence of these events has been to put and end to the 'age of innocence' in the pharmaceutical arena (Pignarre 2003). The public became aware that medicine is not simply about benefits but also carries risks: this shift in perception motivated demands for tighter safety regulation, improved liability regimes and fair compensation. However, compared to other health-related risks, in particular those related to food, the degree and nature of concerns about the safety of pharmaceuticals did not increase significantly until recent years. In the 1980s and 1990s, neither France nor the UK experienced pharmaceutical 'scandals' of the magnitude of the BSE crisis. Public attention focused mainly on containing or eradicating illnesses such as AIDS or cancer, except in rather limited 'anti-drug'/pro-alternative medicine circles. Sporadic concerns were expressed, e.g. about the risks associated with contraceptive or hormone intakes. But they carried much more limited media impact than, for example, the relatively low-risk foot and mouth disease.

Why do the majority of the public seem less concerned by pharmaceutical risks than by other risks? Pharmacists have noticed the high commitment to improve safety standards that followed the pharmaceutical scandals of the 1960s (Pignarre 2003: 47–60). It is usually considered to be an adequate response to the public worries that accompanied the Thalidomide scandal (ibid.). Increasing awareness of the regulator, especially the Federal Drug Administration in the US, drove this movement towards more stringent safety standards. For example, the clinical testing of experimental drugs has become very formalized (Abraham 2002), and normally done in three phases[7] (pre-clinical trials, phase II clinical trials and phase III clinical trials), each phase involving a larger number of people and safeguards. Pharmaceutical products do represent a low risk indeed, and safety control has been reinforced over time. However, this explanation is not entirely satisfactory. As we have underlined, risk perception varies between activities (Slovic 1987). For example, nuclear energy is very safe and combines a small number of incidents and victims and a low degree of public acceptance (Fischhoff 1983). Nuclear energy constitutes an even smaller statistical risk than most pharmaceutical products but it does not mean that people are more confident of living next to a nuclear plant than of taking their daily pills.

One of our working hypotheses is that the acceptance of pharmaceutical products is critically influenced by other key factors, one of them being the 'trust factor'. If we give credit to risk management professionals, who 'in the 1980s and 1990s ... started to find correlations between high public perceived risk and high distrust, and vice versa' (Löfstedt 2005: 9), long-term public acceptance of pharmaceutical risks seems to indicate a situation characterized by trust in the pharmaceutical safety regulation. This is particularly striking when we consider the high level of medicalization of Western societies and the large volume of health-related information, which would otherwise have been a trigger for concern. Although it would be a fascinating topic to explore, our analysis is not about identifying trust factors, but rather uncovering the reasons that led to the breakdown of what could be called a 'pharmaceutical high trust model', and how risk contamination may or may not have taken place as a consequence.

A contribution to the understanding of the perception of vaccines: a comparative study of the MMR and hepatitis B vaccinations in France and the UK

Recent debates on vaccines seem to contradict the rather 'optimistic' picture that risk perception of pharmaceutical products would be intrinsically positive. The MMR crisis in the UK and the hepatitis B crisis in France have both challenged views on the long-term stability of societal acceptance of pharmaceutical products. Looking at the media coverage of these two events leads to considerations about differences and convergences in the 'actual' risk and its perception by the public, and help draw preliminary lessons on the differences/convergences in the acceptability of vaccines.

In the 1980s scientists had been warning the Department of Health in the UK that the measles epidemics that periodically swept the country could only get worse and, in a number of cases, would cause deafness and sight loss or even kill children. The single measles vaccination used at that time simply was not working, with take-up rates too low to create the crucial 'herd immunity' of 95 per cent necessary to eradicate measles. In 1987, 11,000 families took part in clinical trials of MMR with no apparent ill-effects. As a result take-up of MMR increased rapidly, especially in the Nordic countries. The Joint Committee on Vaccination and Immunization suggested that combined vaccination would help increase participation. In 1998, Eurosurveillance reported a coverage level of 92 per cent, putting the UK in a high coverage category, only behind Finland (98 per cent), Sweden (97 per cent) and the Netherlands (94 per cent), but above most other European countries, including France (83 per cent) or Germany (no surveillance system).

A paper published in 1998 in the *Lancet* (Wakefield et al. 1998)[8] was the trigger for the MMR crisis. It suggested a link between autism – a developmental disorder, which often arises in the first few years of life – and inflammatory bowel disease. Importantly, what this article did *not* do was to suggest that there was any evidence the MMR vaccine was connected to either of these problems. It said that studies were 'underway' to pursue this theory. After all, the findings were based on a study of just twelve children, which indicated an obvious lack of epidemiological reliability. The exact content of this article, which was later retracted by most of its co-authors, was of little impact on the controversy and not brought to the attention of the public in much detail before a later stage of the controversy. In fact, the press coverage of a possible link between autism and vaccination started with an article in the *Daily Mail* of 24 January 1998 relating the tragic story of Rochelle, whose son Matthew had begun suffering from behavioural change after he had received the MMR jab. As Rochelle herself described it: 'A mother asked me if he had received his MMR jab, because she'd heard there might be a link with autism. It was like someone had hit me. All the pieces fell into place. I cried.' This story was followed by a number of articles, especially in the *Daily Mirror* and *Guardian*, reporting and discussing concerns about possible links between MMR and autism. Similar stories were reported, and, since then, regular coverage by both tabloid and the broadsheet British press has continued from time to time. For the most part progress does not seem to have been achieved in understanding what could have gone wrong and the same arguments were used again and again. In press statements following the publication of the *Lancet* article, the views of two of the main authors of the paper provided the framework of the discussion. The press reported that one of the researchers, Dr Andrew Wakefield, was suggesting that single vaccines should be offered to parents rather than the combined jab, and this was interpreted as a sign of a link between the MMR vaccine and autism, despite the fact that Wakefield never made this point explicitly. However, another scientist who participated in the research on MMR, Dr Murch, suggested parents should stick with the single jab, which was interpreted as a sign that MMR might not be 'dangerous' after all.

Despite worries reported from time to time, scientific research has provided constant reassurance that the link between MMR and autism is highly improbable.[9] Recently, a study was performed on a reliable sample of 30,000 children in Yokohama, Japan, by researchers from the Yokohama Rehabilitation Centre and the Institute of Psychiatry, London. It was given much publicity and 'has provided powerful evidence that the injection is not linked to the mental disorder'.[10] However, despite the growing flow of reassuring information which led to this happy conclusion, little progress seems to have been made in resolving the public 'controversy' about MMR. A tangible outcome is the continuing decline of MMR vaccine takeup in the UK: it fell to a record low between July and September 2003 with a coverage rate of just 84.2 per cent (the recommended rate being 95 per cent). MMR vaccination has since continued to fall and is now reported to be just over 60 per cent.[11] The increasing demand for single jabs creates a number of issues, especially the lack of available single vaccines, because the industry did not anticipate this. For example, the Department of Health confirmed that 13,000 children received the single vaccine in 2002.

Why has it been so difficult to end the controversy? More than a 'crisis' with a clear beginning and a clear end, this coverage of the relationship between MMR and autism seems to be sustained by the fact that it is offering an ongoing platform of public discussion on vaccine-related risks. The vivid nature of online discussion forums is a good sign of wide public interest for the issue. To understand this we would suggest a debate about vaccines in a context of high command and control strategies: the UK health system is public and vaccination is *de facto* compulsory and driven by public health priorities. Some citizens who may have concerns about the benefits of vaccination may not be reassured by this top-down decision system and therefore would be very receptive to arguments challenging established views, even if this argument be weak from a scientific point of view.

Second, neither the press nor public decision-makers and politicians seem to have been willing to acknowledge the nature of the societal concern about vaccines. To quote the *Independent*, 'Ministers need to borrow some "marketing genius"'. Public response has been conveyed in three stages. The first tried to back public vaccination policy through generating more scientific evidence that there was no link between the MMR vaccine and autism. The second communication stage denied any link and spread the scientific message. This is illustrated by an impressive number of statements concerning the reliability of the MMR vaccine. A key statement was made in November 2003 by the then Health

Secretary Dr John Reid who said categorically, 'MMR does not increase autism risk'.[12] The third government strategy that emerged after 2003 has apparently fostered a 'shame and blame' approach. It increasingly puts the blame on parents who refuse to allow their children to receive the combined jab and on the biased media. In the autumn of 2003 Prime Minister Blair warned of the 'scaremongering of the media'. He also indicated that his youngest son had had the controversial MMR triple vaccine as he attempted to allay fears over its safety. This strategy, though, could appear to be a dangerous path to follow. Considering the low level of trust in public authorities, how credible would their imprecations be perceived by the public? In addition, the possibility of more 'scientific uncertainty' hitting the news cannot be totally ruled out. The headlines provide an argument for those who would call for a cautious communication approach. The *Sun*, for example, asked, after some reassuring news had been conveyed:[13] 'What on earth are worried parents supposed to make of it?' Its editorial stated: 'The public feud against two of the doctors who originally warned of links to autism and bowel disease is making parents' lives hell.' The *Express* headline the same day was 'Now what do we do?'[14] Finally, the lack of a clear communication strategy combined with an aggressive tone would logically reinforce the politicization of the issue.[15]

At some point the BBC underlined that MMR debates were characterized by 'cries and confusion from many of the newspapers'.[16] This confusion emerged from the news coverage itself, with conflicting messages being sent to parents. Although this confusion is begging to be acknowledged, it did not prevent the media from sending conflicting messages. The BBC itself has added to the confusion, highlighting, for example, that 'a little boy who was accidentally given a double dose of the MMR vaccine is being tested for suspected autism'.[17]

Interestingly, it appears that the MMR scandal has had little impact on France, unlike for example Japan, where the MMR vaccine was banned. In the case of the UK and France risk contamination from country A to country B did not occur. Although the controversy affecting the UK was reported in the French media, events taking place in Britain do not seem to have influenced the public debate in France. We found a very limited number of articles in the French press.[18] This may be due to the fact that, at the peak of the MMR controversy, France was actually confronted with major societal concerns about another vaccine, the hepatitis B vaccine.

Hepatitis B is far less common than measles but its consequences are serious in far more cases. Hepatitis B is a viral illness, which, like AIDS, is transmissible by sexual intercourse and blood. It can lead to liver cirrhosis and cancers in a significant number of cases. Vaccine against hepatitis B has been available since 1982 and, since 1996, the World Health Organization (WHO) has recommended that, like measles, all infants should receive the vaccine as part of routine childhood immunization.[19] In 1994 the French Minister of Health, Philippe Douste-Blazy, launched an ambitious vaccination campaign to eradicate hepatitis B. The purpose was, on the one hand, to advertise this vaccine among the general population, and on the other hand, to target systematically children aged 10–12 (Levy-Bruhl et al. 2002), in order to ensure a high take-up. Although the vaccination would not be 'compulsory' it would be proposed and administered systematically, and would be administered in schools. This policy was at first successfully implemented and the result was a substantial coverage of about 75–80 per cent between 1994 and 1997 (ibid.).

It is interesting to note that this initiative was unusual among developed countries, where there is usually only a limited effort to develop large-scale vaccination against hepatitis B. It is also interesting to note that it was pushed in a context of relatively moderate vaccination levels of the far more common diseases such as measles/mumps/rubella or varicella (Coudeville et al. 1999). As a result of this ambitious campaign, 25 million French people, both adults and children, were reported to have been vaccinated against what was described as 'this new HIV that can be caught from saliva'.

Two years later, in 1996, articles in the press began to suggest a link between the vaccine and cases of multiple sclerosis, autoimmune illnesses and lupus, and questioned the systematic vaccination of children, despite the reassuring outcomes of safety enquiries also being reported.[20] The scare started with a series of articles that could be labelled as developing a 'conspiracy theory'. Many of the articles stressed, almost *en passant*, the scientific uncertainty surrounding the vaccine, and focused instead on the covert influence of the pharmaceutical industry and its impact on the public decision to promote the vaccine.

After repeated newspaper articles stressing the level of distrust[21] and expressions of public concern about the development of multiple sclerosis (MS) among people who had been vaccinated, M. Douste-Blazy's successor, Bernard Kouchner, suspended the schooled-based hepatitis B vaccination programmes in October 1998. This was presented as a precautionary measure allowing time for more research to be done, with results promised within a year or so. In fact, gathering extensive epidemiological evidence would have taken much more than a year, as large-scale vaccination was too recent, and therefore tracking back reliable data would have been impossible. Despite the positive conclusions of the studies, which seemed to confirm the absence of a link between the hepatitis B vaccine and MS,[22] the public authorities preferred to ignore the results and the vaccination campaign in schools never recommenced. Since then, despite persistent voices calling a re-engineering of the vaccination campaign,[23] no renewed state commitment has been observed.

The political resolution of the crisis was much faster than in the case of the MMR, and the French government could at first appear to have taken a sound decision. Its attitude was generally understood in the media as an illustration of a robust application of the precautionary principle, although some organizations opposed the decision, especially ACT UP, an anti-AIDS organization.[24]

In 'externalizing' the crisis, the French government managed to protect itself from the stigma affecting the UK government in the case of the MMR vaccine. But when looked at more carefully, the fast reaction of the French government – bringing an end to the crisis without any attempt to confront public perceptions – seems to have had unexpected and costly consequences: the issue was moved to Parliament and the judiciary.

A proposal for a parliamentary resolution on the hepatitis B vaccine was introduced on 14 February 2001,[25] which finally did not come through. But it summarized the nature of the debate by challenging the annual report of a pharmaceutical company which asserted: 'vaccination is good for health and good for shareholders' (*'la vaccination c'est bon pour la santé, c'est bon pour les actionnaires'*) on the basis that 'profit and health rarely get along well' (*'Le profit et la santé font rarement bon ménage'*). This was of course a potentially destructive public health message.

The consequences of the scandal were more significant for the judiciary. The introduction of compensation claims of people suffering from MS put courts in a difficult position. Winning the legal battle soon emerged as a key target of anti-vaccination groups. According to the American journal *Science* (1998), French attorneys representing 15,000 French citizens filed lawsuits against the government 'accusing it of understating the vaccine's risks and exaggerating the benefits for the average person'. Media reported 'stories' such as this one of a French physician who reportedly collected data on more than 600 people suffering from serious immune and neurological dysfunction following hepatitis B vaccination, many with symptoms resembling MS. Following a long tradition of distrust between the executive and the judiciary,[26] the judiciary refused to endorse the 'no action' policy of the government. The courts considered that they were not competent to assess the risk, and some of the complainants were finally granted compensation on the basis that the government was better informed and that the precautionary measures taken by the minister implied the possibility of a link between the vaccine and the disease. Of course, the opponents of the vaccination took this decision as a victory.[27]

Since the French government decided to stop the vaccination campaign, further large-scale epidemiological surveys have been completed, providing reassuring information about the safety of the hepatitis B vaccine (Sadovnick and Scheifele 2000; Ascherio et al. 2001; Confavreux et al. 2001; De Stefano et al. 2003), and have therefore consolidated the WHO's position. A dissident view from a recent study looking at UK data has somewhat re-engineered the controversy (Hernan et al. 2004), but flows in the selection process of the sample have also been identified[28] (Ritvo et al. 2005).

Conclusion

The two crises have had similar consequence: a precipitous decline in the take-up of the vaccine. One of the key differences is that the UK faces an ongoing political crisis, while the French government has managed to avoid this. However, the French crisis was 'imported' into the legislative and judicial arenas, with destructive and costly implications and at the risk of discrediting the entire medical system. In both cases the event of a serious political backlash in the future cannot be ruled out.

The comparative analysis of the MMR and hepatitis B cases also conveys a series of general lessons:

- Trans-boundary risk contamination, between various issues, has been observed in France and the UK. However, despite very similar challenges and common policy agendas in both countries, there is no evidence that the hepatitis B and MMR cases influenced each other's policies. The hypothesis that common concerns for 'vaccines' may be a cause of future risk contaminations nevertheless remains plausible.
- Although a more detailed study would be necessary to test our assumptions further, there are strong indications that distrust in government played a central role in the controversy; this is particularly important if we consider that in both cases public authorities were heavily involved in the vaccination campaign. High visibility on the one hand, and distrust on the other, may have engineered public scare.
- When it comes to risk perception, the quality of scientific information is not sufficient to back up risk communication. Reassuring

expert messages concerning the safety of the MMR vaccine has not stopped the decline of take-up. Only 'trusted' sources of information are credible in the public's eye, and trust does not equal scientific credibility or experts' views. This militates in favour of a two-way risk communication strategy rather than the translation of experts' knowledge to laymen.
- The potential financial, political and health-related costs of the policy failure are significant and the entire vaccine policy of a country could be affected. Vaccinations take-up is declining in both countries, with potential implications for public health.
- It has been observed that vaccine acceptance is a particular challenge when the prevalence of a disease is low (Slovic 1987; Ritvo et al. 2005). The comparative analysis of the MMR and hepatitis B vaccines may indicate that this is taking place independently of the severity of the disease.

In an article published in 2001 Baruch and Ilya Fischhoff analysed the mechanisms of societal attitudes to risk from the observation of contemporary developments. They noted that in a number of areas, including for example the production of GM food in Europe, specific technologies seem to be increasingly perceived as 'taboos' and are being 'stigmatized', and are therefore not eligible for trade-offs of potential risks and benefits. These stigmatized activities are considered to be intrinsically unacceptable. Interestingly, they observed the absence of a clear relation between activities that the public considered to be unacceptable, and the risk estimates made by the experts (Fischhoff and Fischhoff 2001).

The analysis of the MMR case indicates that poor communication is no solution and the hepatitis B example shows that 'no action' policies are not satisfactory either. Concerns must be addressed thoroughly, which implies considering sound risk communication as an option to avoid stigmatization. Avoiding the stigmatization of vaccines through an informed debate taking on board the lessons of 50 years of risk communication research will be a particularly important challenge for the future. We should keep in mind that immunization programmes have been credited with the elimination of infectious diseases including smallpox, and the near-eradication of polio (Plotkin and Plotkin 1999). 'Herd immunity presents bioethical challenges because individuals do not just benefit from their own vaccination, but from the vaccination of others. Thus individuals who refuse vaccination can benefit from the vaccinations others undergo. But their refusal to be vaccinated can also endanger others, including those who have been vaccinated' (Ritvo et al. 2005). In this difficult context of vaccines, defining a risk acceptability criterion, i.e. creating the right 'workable compromise' among citizens, may seem necessary. We do not believe that this 'workable compromise', which would imply shared decisions about the level and nature of tolerable risks, could be achieved in France or the UK without a decisive effort. This constitutes a powerful transnational lesson: vaccines could well be in the vanguard of a deeper distrust in the safety of pharmaceutical products to come, as the recent, high-profile recalls of Bayol and Vioxx could well indicate.

Notes

1 Members of the OECD will be used as a conventional grouping for 'Western countries'.
2 The journal *Risk Analysis*, published by the Society for Risk Analysis gives a clear account of these ongoing scientific developments.
3 Renn (1991) defined risk amplification as '[being] based on the thesis that events pertaining to hazards interact with psychological, social, institutional and cultural processes in ways that can heighten or attenuate individual and social perceptions of risk and shape risk behaviour'. Social amplification does not need the actual occurrence of a risk-related event to exist and it can be observed independently of the substantive or the hypothetical character of a risk (in Kasperson et al. 1988). Transmitters play an essential role in risk amplification, the mass media being the most usual risk transmitter.
4 See, for example, the debates on the Social Amplification of Risk at the 2001 Hazard forum: http://www.hazardsforum.co.uk/SAR per cent20Report.htm.
5 E. Millstone, The Painful Lessons of BSE, *Financial Times*, 6 October 2000.
6 See, for example, Ragnar Löfstedt, Time to Sow Seeds of GM Harmony, *Independent*, 25 August 2002; Ragnar Löfstedt, So Should We be Scared of Salmon? *Independent*, 10 February 2004.
7 A fourth surveillance phase is also added at the post-marketing stage.
8 See also http://thelancet.com.
9 Research backs claims of MMR safety. *Financial Times*, 12 June 2002, p. 4. Study Finds no Evidence of MMR and Autism Link, *Daily Telegraph*, 12 June 2002, p. 8; No Evidence of Link between MMR and Autism, Doctors Find, *Independent*, 12 June 2002, p. 7; Research Analysis Rules out MMR Risk, *The Times*, 12 June 2002, p. 8; Jab Gets Safety All-clear, *Sun*, 12 June 2002, p. 20; MMR 'is Safe'. *Daily Express*, 12 June 2002, p. 26; The MMR Jab is Safe. *Daily Mirror*, 12 June 2002, p. 17.
10 Study Finds MMR Jabs not Linked to Cases of Autism, *Independent*, 3 March 2005.
11 BBC, 3 March 2005: 'No Link' between MMR and Autism
12 BBC, 2 November 2003.
13 *Sun*, 1 November 2003.
14 *Express*, 1November 2001.

15 For example, on 6 February 2002, the Conservative Shadow Health spokesman, Liam Fox, declared: 'Whether rightly or wrongly, the public has lost confidence in the government's policy on MMR'. Source: BBC website.
16 BBC, 1 November 2003.
17 BBC, 17 October 2002.
18 Coverage came mostly from specialized media and were rather positive towards the vaccine: http://www.e-sante.fr/magazine/article.asp?idArticle=5238&idRubrique=4
19 http://www.who.int/child-adolescent-health/New_Publications/NUTRITION/updt-22.htm; WHO Expanded Programme on Immunisation: Measles and Hepatitis B. http://www.wpro.who.int/pdf/epi/tag14 _report_draft
20 VSD, 14 and 20 November 1996, *Vaccin anti-hépatite B Prochain lancement d'une enquête officielle* (Anti-hepatits B vaccine, soon under public enquiry); *Alternative Santé*, Mai 1997 Les Hépatites B et C et les autres [Hepatitis B and C and others], *Le Parisien*, 21 January 1998, 175,000 enfants concernés – revelations sur un vaccin dangereux [175,000 Children concerns – revelations on a dangerous vaccine]. *Pelerin*, 6 and 13 February 1998, Hepatite B Questions sur un vaccin [Hepatitis B, questions about a vaccine].
21 See, for example, a series of articles by the magazine *Alternative Santé*, a well-known periodical promoting alternative medicine offered a series of articles
22 See *Le Monde* 12 March 1999, Confirmation de l'efficacité du vaccine contre l'hépatite B [Confirmation of the effectiveness of the Hepatitis B vaccine].
23 See *Le Monde*, 23 September 2003, Hépatite B: une priorité sanitaire pour les enfants de moins de 2 ans [Hepatitis B, a health priority for children under 2 years].
24 http://www.actupparis.org/bkment/j-19onverra.html.
25 Proposal for the creation of a parliamentary commission on development of the vaccination campaign, 14 February 2001. Assemblée Nationale, Nb. 2930. The commission was never established.
26 As a result of this tradition, which goes back to the French Revolution, the judiciary is not made a distinct 'power' comparable to the executive or the legislative, but an 'authority', which is to some extent subordinated. According to the 1958 Constitution its independence is *guaranteed* by the President of the French Republic (arts. 64-66 of the French Constitution).
27 http://www.medecines-douces.com/impatient/279juin01/vaccin.htm.
28 World Health Organization. http://www.who.int/vaccine_safety.

References

Abraham, J. (2002) Pharmaceutical Industry as a Political Player, *Lancet*, 360: 1498–1502.
Ascherio, A. et al. (2001) Hepatitis B Vaccination and the Risk of Multiple Sclerosis, *New England Journal Medicine*, 344(5): 327–332.
Ball, D. J and S. Boehmer-Christiansen (2002) *Understanding and Responding to Societal Concerns* (London: HSE Books).
Bostrom, A. (1999) Who Calls the Shots? in George Cvetkovich and Ragnar Löfstedt (eds.) *Credible Vaccine Risk Communication, Social Trust and Management of Risk* (London: Earthscan).

Confavreux, C., S. Suissa, P. Saddier, B. Bourdes and S.Vukusic (2001) Vaccines in Multiple Sclerosis Study Group. Vaccinations and the Risk of Relapse in Multiple Sclerosis, *New England Journal Medicine*, 344(5): 319–326.

Coudeville, L., F. Paree, T. Lebrun and J-C. Sailly (1999) The Value of Varicella Vaccination in Healthy Children: Cost-benefit Analysis of the Situation in France, *Vaccine*, January, 17(2): 142–151.

Covello V., P. Sandman and P. Slovic (1988) *Risk Communication, Risk Statistics, and Risk Comparison: A Manual for Plant Managers* (Washington, DC: Chemical Manufacturers Association).

DeStefano, F. et al. (2003) Vaccinations and Risk of Central Nervous System Demyelinating Diseases in Adults, *Archives of Neurology*, 60(4): 504–509.

Fischhoff, B. (1983) Acceptable Risk: The Case of Nuclear Power, *Journal of Policy Analysis and Management*, 2, 559–575.

Fischhoff, B. (1994) Acceptable Risk: A Conceptual Proposal. Risk: Health, *Safety & Environment*, 1, 1–28.

Fischhoff, B. (1995) Risk Perception and Communication Unplugged: Twenty Years of Process, *Risk Analysis*, 15(2): 137–145.

Fischhoff, B. and Fischhoff (2001) Publics' Opinion about Biotechnology, *AgBioForum*, 4(3–4).

Fischhoff, B., A. Bostrom and M. Jacobs-Quadrel (1993) Risk Perception and Communication, *Annual Review of Public Health*, 14: 183–203.

Fischhoff, B., P. Slovic, S. Lichtenstein, S. Read and B. Combs (1978) How Safe is Safe Enough? A Psychometric Study of Attitudes towards Technological Risks and Benefits, *Policy Sciences*, 8: 127–152. Reprinted in P. Slovic (ed.) *The Perception of Risk* (London: Earthscan, 2001).

Giddens, Anthony (1991) *Modernity and Self-identity. Self and Society in the Late Modern Age* (Cambridge: Polity Press).

Hernan, M.A., S. S. Jick, M. J. Olek and H. Jick (2004) Recombinant Hepatitis B Vaccine and the Risk of Multiple Sclerosis: A Prospective Study, *Neurology* 63(5): 838–842.

Kasperson, R. E. et al. (1988) Social Amplification of Risk: A Conceptual Framework, *Risk Analysis*, 8: 177–187.

Levy-Bruhl, D. (2004) *Impact of Hepatitis B Vaccine Safety Issues on Vaccination Strategies and Their Implementation in France. Viral Hepatitis Prevention Board. Meeting on Prevention and Control of Viral Hepatitis in France: Lessons Learnt and the Way Forward*, 18–19 November.

Levy-Bruhl, D., J. C. Desenclos, I. Rebiere and J. Drucker (2002) Central Demyelinating Disorders and Hepatitis B Vaccination: A Risk-benefit Approach for Pre-adolescent Vaccination in France, *Vaccine*, 20(16): 2065–2071.

Löfstedt, R. E. (2003) The Precautionary Principle: Risk, Regulation and Politics, *Trans IChemE*, 81, part B, January: 36–43.

Löfstedt, R. (2005) *Risk Management in Post-Trust Societies* (Basingstoke and New York: Palgrave Macmillan).

Morgan, M. G., M. Henrion with Small M. (1990) *Uncertainty: A Guide to Dealing with Uncertainty in Quantitative Risk and Policy Analysis* (New York: Cambridge University Press).

Morgan, M. G. and Fischhoff, B. (2001) *Risk Communication* (New York: Cambridge University Press).

Renn, O. (1991) Risk Communication and Social Amplification of Risk, in R. E. Kasperson and P. M. Stallen (eds.) *Communicating Risks to the Public: International Perspectives* (Amsterdam: Kluwer).
Pignarre, P. (2003) *Le Grand secret de l'industrie pharmaceutique* (Paris: La Decouverte).
Plotkin, S. L. and S. A. Plotkin (1999) A Short History of Vaccination, in S. A. Plotkin and W. A. Orenstein (eds.) *Vaccines* (Philadelphia: Saunders).
Powell, D. and W. Leiss (1997) *Mad Cows and Mother's Milk, The Peril of Poor Risk Communication* (Ontario: McGill-Queens University Press).
Ritvo, P., K. Wilson, D. Willms and R. Upshur (2006) Vaccines in the Public Eye, *Nature Medicine*.
Sadovnick, A. D. and D. W. Scheifele (2000) School-based Hepatitis B Vaccination Programme and Adolescent Multiple Sclerosis, *Lancet*, 355(9203): 549–550.
Slovic, P. (1987) Perception of Risk, *Science*, 236: 280.
Slovic, P. (1998) The Earthscan Reader, in R. Lofstedt and L. Frewer (eds.), *Risk and Modern Society* (London: Earthscan).
Slovic, P., B. Fischhoff and S. Lichtenstein (1981) Perception and Acceptability of Risk from Energy Systems, *Advances in Environmental Psychology*, 3 (Hillsdale, NJ: Erlbaum).
Starr, C. (1969). Social Benefit versus Technological Risk: What is Our Society Willing to Pay for Safety? *Science*, 165: 1232–1238.
Vogel, D. (2003) The Politics of Risk Regulation in Europe and the United States, The *Yearbook of European Environmental Law*, 3.
Wakefield A. J. et al. (1998) 'Ileallymphoid-nodular Hyperplasia, Non-specific Colitis, and Pervasive Developmental Disorder in Children', *Lancet*, 351: 637–642.
Wiener, J. and Rogers, M. (2002) Comparing Precaution in the United States and Europe, *Journal of Risk Research*, 5(4): 317–349.
Wildavsky, A and K. Drake (1990) Theories of Risk Perception: Who Fears What and Why? *Dædalus*, 119(4): 41–60.

11
Risk, Precaution and the Media: The 'Story' of Mobile Phone Health Risks in the UK

Adam Burgess

Introduction

This chapter surveys the role of the media in generating health concerns about mobile phones in the UK. Despite the near-universal public embrace of this technology, the British media has adopted a concerted campaigning stance around possible health risks from mobile phone electromagnetic emissions. The chapter examines the media's influence on campaigning and in stimulating 'precautionary' government action. It also highlights the role of the Internet in the problematizing of mobile phones. The chapter reviews the story of the mobile phone panic in the context of suggestions that with the decline of more traditional influences, the media now play a more influential social role, even becoming 'civil society' itself. It concludes that the media were a crucial influence in galvanizing government reaction, independently of any widespread public demands.

'Public' health concern about mobile phones?

'New' and 'old' media alike have been instrumental in establishing a high profile for health concerns about the mobile phone – an instrument itself, of course, at the cutting edge of the new media through the anticipated mobile Internet revolution. Beginning in the United States in 1992 and in the UK around five years later, the media have prominently featured a wide range of potential hazards associated with mobile phones and base stations (commonly known as 'masts'). Media coverage of the mobile phone has consequently been overwhelmingly negative. The Independent Expert Group on Mobile Phones (IEGMP, also known as the Stewart Group) reviewed 641 cuttings from newspapers published in the UK between January 1999 and February 2000, and examined the content of 76 radio and television programmes.[1] Seventy-nine per cent of the media reports alleged that mobile phones and base stations were causing adverse health effects, whereas only 9 per cent concluded that there was too little rigorous scientific evidence to arrive at any conclusion, or reported no adverse effect (IEGMP, 2000). As is elaborated below, a number of British newspapers adopted an explicit campaigning stance on the issue. The concluding report of the Stewart Group itself gave rise to widespread newspaper comment on the possibility of health risks from mobile phones based on the report's 'precautionary' proposition that if there were a hazard, children might be especially vulnerable.[2]

Mobile health concerns are becoming an issue in an increasing number of countries.[3] Electromagnetic (EMF) emissions from these devices are at the centre of concern, most dramatically with the allegation that they could cause cancer. Lawsuits based on this continue to dog the mobile phone industry in the US. Along with the many other problems confronting the realization of the 3G mobile Internet network, the construction of the thousands more masts necessary is now threatened in the UK. An influential national campaign group backed by MPs, Mast Action UK, is successfully promoting concerns about mobile mast sites, and organizing protests. Although having accepting billions of pounds in licence fees for the 3G system which will require thousands more base stations, the government is imposing tighter restrictions on mast erection.[4]

Despite widespread coverage and political reaction, however, this problematization of mobile telephony appears to have been of little 'public' consequence. Mobile usage has continued its extraordinary increase, with around two-thirds of Britons now owning a mobile. This is not to suggest that mobile health concerns have been entirely rejected. Witness the – albeit temporary – health-based popularity of 'hands-free' kits. Nagging parental concerns about children's usage appear to persist. Nevertheless, it can be argued that the risk has effectively been outweighed by the considerable convenience and utility of the mobile. Despite official endorsement of anxieties about children's usage by the Stewart Enquiry, the ability of mobiles to keep parents (potentially) in touch with their children seems to have put intangible concerns about electromagnetism to one side. The apparent lack of demonstrated public concern is curiously at odds with both an internationally unprecedented media campaign on mobile EMF, and government reaction that has recognized 'public' concerns.

This chapter reviews the important role of the media in the 'story' of the mobile phone panic in the UK. It provides a specific examination of the reciprocal interplay between media and society (Croteau and Hoynes 1999). Recent work asks whether political culture is being transformed by the new media (Axford and Huggins 2000). Here we examine how heightened perceptions of risk from new media technologies might be affecting such a transformation; how contemporary political culture might be limiting the impact of the 'third generation' of communication. The suggested centrality of the media in the construction of technological risk perception is worth closer examination, particularly where it appears that it is the 'old media' that is questioning the implications of the new. Through a contemporary case study it may be possible to shed light on the perennial question of whether the media play a creative role in problem construction, or are merely messengers of emerging social concerns. Themes of sender, message and direct and indirect effects in the research tradition are most usefully explored in McQuail's important work (2000). A more specific issue is the relationship between public, government and the media.

A core concern of contemporary social theory is the transformation of the public sphere. It is noted that the public sphere is no longer the site where the competing claims of organized networks and bodies now play themselves out (Habermas 1989; Melucci 1989; Giddens 1990; Donati 1997). One consequence of the decline of the public sphere is the new importance of the media. The media are seen to have become increasingly autonomous from politics and the state. It is even argued that the media have become the public sphere itself, with the decline of other content to public life (Diani and Donati 1999: 28). As 'civil society' has shrunk, so the media have bypassed more traditional democratic mechanisms, effectively becoming contemporary 'civil society' itself. The issue of the media's own direct influence on governance has arguably become more separate from that of public 'interference' than ever before.

Issues of contemporary media influence, political impact and the relationship to public perception have a strong bearing on the case of mobile EMF risk. A little noticed feature of the government-sponsored enquiry, the IEGMP, is the discrepancy between the recognition of the objective lack of serious concern about health risks among the general population and the imperative to be responsive to what are described as 'public' perceptions. In the first of the general conclusions considering public perceptions it is stated that: 'The continuing rapid growth in the use of conventional mobile phones ... indicates that most people do not consider the possibility of adverse health effects to be a major issue' (IEGMP, 2000: 26). The enquiry made clear that it was aware that the complaints it had heard – particularly at the five public meetings organized to gather opinions – were not representative of the population at large. Overall, 'a small minority of highly proactive individuals have led the public debate on possible health hazards' (IEGMP 2000: 24). Yet, despite acknowledgement of the limited evidence of public concern, the IEGMP declared it important for all institutional parties to address these minority anxieties; the authors 'further recommend that national and local government, industry and the consumer should all become actively involved in addressing concerns about the possible health effects of mobile phones' (IEGMP 2000: 113).

Perhaps the most interesting aspect of the mobile phone story is how, through the media, an originally marginal and scientifically unverifiable concern has subsequently become an important focus for government policy and action. The government commissioned a large-scale enquiry that subsequently endorsed precautionary measures on the basis that there was some evidence of subtle biological effects, even if there was no evidence to confirm that these effects were negative. In the shadow of the BSE experience and in what has been tellingly dubbed the 'risk society' it appears to be considered necessary to respond to even the most marginal of risk concerns (Beck 1992). Government-sponsored action on the issue continues into 2001, with a new committee disbursing several millions of pounds to conduct further scientific, sociological and psychological research into mobile phone risk.

The discrepancy between public and government reaction alone makes the mobile phone risk an interesting case study. Media promotion of the risk is also an excellent subject matter because it is a relatively 'pure' story. Concern about the EMF emitted by mobiles is a 'phantom risk' that can be traced as an idea with relatively little interference from real events and pressures (Brauner 1996; Foster et al. 1999). Despite ongoing research and extensive reviews of existing studies, there is still no consistent evidence of any negative health impact from mobile phone emissions, let alone a serious one. According to the latest WHO fact sheet (2000): 'present scientific information does not indicate the need for any special precautions for use of mobile phones'. The most recent and authoritative research confirms in relation to the chief focus of concern – brain cancer – that there is no greater risk of tumours among cell phone users (Muscat et al. 2000; Inskip et al. 2001). Long-term effects remains a source of concern for some, but such a possibility both remains hypothetical and assumes the unlikely possibility that current technology remains static (hence making long-term exposure from the same devices possible). Continuing uncertainty rests principally on the scientific impossibility of proving that anything, EMF included, is 'safe'.

This is not to say that some people have claimed to be affected. Indeed, the whole story of the mobile phone 'panic' began with a 1992 lawsuit in the US by a man claiming his wife's death from cancer was caused by mobile EMF. Every major newspaper ran at least one item over the fortnight following the breaking of the David Reynard case. An industry poll conducted shortly after Reynard's appearance on *Larry King Live* found that half of all Americans knew about the lawsuit (*Microwave News* January/February 1993), and shares in mobile phone companies plummeted. As a result of the chain of events unleashed by the Reynard case, the Federal Drug Administration announced an advisory committee on mobile phones. And it is not only in American lawsuits that actual harm is alleged. The moral driving force behind the now national campaign against mobile phone masts in the UK is the effect on children alleged by some campaigners. Yet most campaigns, and certainly the important political concern about mobile phones, are based on the possibility of harm; or conversely that it cannot be ruled out. It is the potential rather than demonstrated effect upon children that is at the centre of concern. This precautionary response is acknowledged to be a result of the general political and cultural impulse towards precaution in 'post-BSE' Britain, rather than anything specifically related to the effects of mobile EMF. In this sense the reaction against mobiles cannot even meaningfully be characterized as (typical British) 'technophobia'. It is not based on suspicion of the unknown and invisible, so much as the 'better safe than sorry' precautionary culture that suggests even probably unfounded risks might just as well be avoided.

Mobile EMF concern has not emerged from growing evidence of harm to individuals. The influential early stories in particular (as is illustrated below) were based on hypothetical harm to humans derived from one-off scientific experiments on animals. As it is not principally based on widespread claims of real injury, the idea of potential harm from mobile EMF must have an identifiable source and method of transmission. Certainly, it has not been reported simply because 'it happened'. A significant proportion of UK coverage, for example, has been led by the claims of a single academic, Dr Gerard Hyland, senior lecturer in theoretical physics at the University of Warwick. Hyland has been given significant space to expound his theory of possible risk. His ideas are not based on extensive new research, but on a longstanding hypothesis about the possibility of 'non-ionizing' (i.e. too weak to have a heating effect) radiation interacting with some humans (Hyland 2000). This possibility is based on analogies such as with the effect of strobe lighting on epileptics. The fortunes of such theoretical claims can be followed as a story relatively independent of important real developments which cannot be attributed to any active media role, such as a mass cluster of cancers among mobile users.

At the grassroots level these health concerns appear to have at least partly arisen through interaction with a UK media that appears predisposed to highlight possible risks in all walks of everyday life.

'Old' and 'new' media galvanize health-based opposition

Although concerns about mobile masts developed after anxieties about handsets, it is useful to begin with a brief examination of the dynamics of the local protest that emerged in the UK from late 1998. Although media problematization of handsets is chronologically prior in the mobile risk story, first outlining the reaction against masts helps clarify that the media have a more active and integrated role than simply (over)reporting 'scare stories' at the national and international levels. Some evidence from my own research suggests that the media have played an important role in energizing and generalizing local concerns about 'inappropriately' sited masts.[5] Most importantly they have played a role in elevating specific health concerns about nearby masts over and above other factors such as issues of local democracy, property and aesthetic concerns. The media always have an ability to generalize from the specific to the general; and in the 'risk society' this takes the form of promoting health concerns over more mundane local issues.

The focus for active concern about mobile telephony is mobile masts, principally when erected near an individual's residential property or, even more potently, their children's school. In specifically health terms, such a reaction is something of a curiosity. The Stewart Report itself explained that even if the possibility of a health risk associated with mobile EMF were to be accepted, the masts pose almost no potential risk compared to handsets. The likelihood of weak emissions from these structures having any sustained impact is remote and hardly comparable with holding a device against one's head for prolonged periods. The scientific illogicality of reaction to masts rather than handsets underlines the fact that reaction is propelled by a dynamic driven by social influences rather than scientific knowledge or actual harm. Often, protest began simply as a reaction to the appearance of an ugly structure and was not informed by health issues.

The often unannounced appearance of these structures next to people's property has understandably caused a reaction. Interviews with numerous protestors against masts indicate that reaction is dominated by indignation about a lack of consultation along with other competing factors. Generally, it is only in relation to school masts that more diffuse health-based anxieties were relatively spontaneous. Even in these cases, however, suspicion of anything unfamiliar appearing close to their children was evidently not informed by any sort of specific knowledge about how these structures might cause harm. The extent to which health anxieties appeared was also dependent on how developed the media profile was at any one time. And it was often only through the process of 'taking things further' in terms of finding out more information and publicizing their plight that the possibility of harmful health effects was introduced and often came to override other factors.

The 'old' media played an often instrumental role in creating campaigning momentum and consolidating the health aspect of reaction to masts. Margaret's successful campaign began by going directly to the local papers to publicize the way in which the offending mobile operator had put up a mast without consulting local residents. Important to the launch of Peter's campaign was the fact that his neighbour was a regional television news producer, so he had the inspiration and potential access through which he could gain coverage of his grievance. Sometimes it was the media that effectively made the story for themselves. Tony and Jo's campaign was effectively initiated by *BBC Online*, when they began interviewing parents and the headmistress about plans (at that time unknown to the parents) for a mast on a school. The campaign 'snowballed' after this report ('So these things can harm my child, can they?'), with *Anglia News* wanting a story, as well as local radio stations and newspapers. An even more common pattern was of individuals hearing about other complaints through the media. This allowed them to generalize from their own experiences and provided a direct link to others who would otherwise have remained unknown. Sue heard on the radio that Liverpool Council were ordering an operator to take down a mast. She managed to take her own efforts further after establishing contact with the council. Shortly after the mast went up near Mary's property, she saw an article in the *Express* by the campaigning journalist Cathy Moran. She called her and was then put in contact with others. Similarly, an article in the *Daily Mail* was an important catalyst for Tony and Jo, and led them to the chief UK scientific voice of alarm, Dr Gerard Hyland.

An important consequence of media promotion of campaigns was that it enabled individuals to identify one another and create campaign links. Establishing contact with others confronting similar circumstances was central to moving from initial outrage to an organized campaign. An organizer from an influential campaign fulcrum, the Liberal Democrats on Kent County Council, recalls that 'many people telephone thinking, "I really feel as if I'm the only person fighting ..." But they quickly realize that it's not the case when I tell them we've had 250 calls ...'. Campaign links were established through the television and newspaper media. Following their coverage on *BBC Online*, Tony and Jo were contacted by another important early organizer. In turn, he put them in contact with others who were similarly contesting the erection of a mast on a local fire station. Jan chanced upon a television news report on Clive's campaign. Getting his number from Carlton Television, she linked up with him. Interestingly, the 'old' media appear to have played a potentially new role as campaign organizers, or at least facilitators in the mobile case.

The 'new' media were also important in establishing campaign links. Roy summed up the sense of campaigners being part of a 'virtual community' in a letter:

> I'm so pleased that we were able to make contact, I believe that campaigns such as ours in joining forces, will achieve far greater overall success than if we were to remain in isolation. Our members are constantly trawling the Internet for likely contacts, much of our success with local authorities has come from this source.

The Internet provided access to material on the alleged health effects of electromagnetism. Margaret's witness statement to the Stewart Group details that her first reaction against the mast was: 'I subsequently updated our computer system, gained access to the Internet and started my investigations.'[6] Among other things she was then able to establish links with individuals from an already established American anti-EMF group. Once part of the loop of activists, individuals were often sent the latest research findings found on the Internet by others. Without the Internet, it would have been difficult to make such materials accessible to ordinary campaigners. This factor is very important because typically it was only through the experience of finding 'alternative' scientific materials on the Internet that health concerns began to acquire a consistent focus. This raises important questions about the Internet as a source of (particularly scientific) information. Any member of the public has been able to gain access, through the Internet, to a range of data, news reports and opinion that might otherwise be unavailable. But there is no process of quality control – a fact particularly important in scientific matters, as they often require considerable interpretive expertise. Good science is dependent on a rigorous process of peer review; without such a process there is no guarantee that research is valid. But the stringent requirements of peer review are no barrier to posting material on the Internet. For the 'non-scientific' campaigners, the importance and implications of this absence of quality control may not be fully recognized. Instead, papers by those without expertise have equal weight with the published results of elaborate and painstaking research carried out by teams of internationally renowned scientists.

One campaigner, perhaps overstating the case somewhat, suggested that, 'this has built up into a movement, which has reached global proportions through the aid of the Internet and email facility'. If the scale of the movement is somewhat exaggerated, there is little doubt as to the important role that the Internet and email played in taking initial outrage further. Brian searched the Internet for similar situations and found out about the various campaigns worldwide. Indeed, what often distinguished those who became involved was access to a computer. The first UK campaign was sustained by the webpage Roy set up with the help of his son. Many of the early activists had a presence on the web and this partially determined the significant role they were to play in the campaigning network. Peter believes that a book could be written on the role of the Internet in contemporary campaigning. For Dahleen: 'it highlighted just how useful the Internet can be'. In both the two early Scottish cases that led Friends of the Earth (FOE) Scotland to its important involvement, the two individuals turned to the web. FoE Scotland itself, once it had decided to lead the campaign, found its information from the web.

The media's campaign against mobile phone health risks

The backdrop to anti-mast campaigning acquiring momentum was the increasing prominence given to negative stories about the effects, at least initially, of mainly mobile handsets rather than masts. Health concerns associated with mobile telephony first became public in the UK following reports about EC concern in 1995.[7] Two notable landmarks were an edition of the BBC television programme *Watchdog Healthcheck*, and a report in the *Sunday Times* with the now infamous headline claiming the possibility that: 'Mobile phones cook your brain' (14 April 1996). The story was by Jonathan Leake who, five years later as the *Sunday Times* science editor, continued to draw attention to possible health risks despite studies suggesting there was no evidence of harm from mobile EMF.[8] The principal focus for the original story was advanced publicity for a study suggesting that rat's DNA molecules could be split by mobile phone radiation. In addition, EC announcements that the Commission intended to publish new, 'safer' emission guidelines and fund new research into mobile phone EMF provided substance for this and other early mobile phone stories. Leake followed up his article the following week (*Sunday Times*, 21 April 1996), adding that one of the scientists who conducted the experiment was now recommending the use of vitamins to combat radiation. The new story was prompted by further inferential suggestion of harm, with the launch of the 'Microshield' microwave protection device (which generated a number of articles in other newspapers), and news of what was purported to be the first person in the UK to begin legal action alleging that mobiles had damaged their health. By the end of April 1996, the idea of a mobile phone health 'panic' began to be referred to casually in the press (*Daily Telegraph*, 26 April 1996; *Sunday Telegraph*, 2 June 1996). By July 'alarmist headlines about mobile phones "cooking" the brain' were picked up in the *Irish Times* (22 July 1996). The *Sunday Times* headline claim that phones might 'cook' the brain was reproduced more widely, and reported as part of a series of 'scare stories' that had prompted the EC to commission new research (*Observer*, 1 September 1996).

The BBC's *Watchdog Healthcheck* (3 June 1996) was more directly influential in promoting awareness of the issue. The programme was trailed and reviewed in most of the broadsheet and specialist press, focusing – like the *Sunday Times* – on the rat DNA molecule experiment (*Independent*, 3 June 1996; *The Times*, 3 June 1996). The programme, and newspaper reviews, pointed out that the scientists reporting ill effects either now limited their mobile phone use or avoided it altogether. Two weeks later, with advance publicity from the *Observer* (16 June 1996), *Watchdog Healthcheck* followed up on its original report (17 June 1996). The programme now claimed that individuals planned to pursue legal action and that two local councils were expressing concern about possible health effects on their employees. At the end of 1996, the influence of the *Watchdog* report within the media continued to be acknowledged, with suggestions that it had led to the commissioning of numerous research programmes (*Guardian*, 14 November 1996).

Sustained media promotion of the issue, principally through the newspapers, did not begin in earnest until the autumn of 1997, however. It was at this time that mobile phone health stories were taken up by the tabloids, and became a regular news feature. This first wave of alarmist reports reached its peak in the summer of 1998. Even *Microwave News*, the pre-eminent American monitor of developments in the field and a promoter of concern about microwave health effects in its own right, described a 'bad summer for mobile phones in England. The British press has been blaming them for everything from hypertension to miscarriages ...' (*Microwave News*, July/August 1998). The tabloids took up the issue as an explicit campaigning focus, and vied with each other to present themselves as the champions of public concern about health risks associated with mobile phones. The *Sun* was the first to adopt a campaigning stance, with its 'menace of our mobiles' slogan in the autumn of 1998. But the later boast of the *Sunday Mirror* that it was they who initially 'led the way' with their 'mobile phone watch' campaign (11 April 1999) was not misleading, as the *Sun* did not sustain its interest.

The *Sunday Mirror*'s campaign made a wide range of health claims: that mobile phones simply make you ill (16 May 1999) – especially if you wear glasses (30 May 1999). They can affect your liver and kidneys (11 July 1999); more generally, that 'a mobile puts years on you' (17 October 1999). Introducing a sinister conspiratorial theme, they claimed that: 'experts knew 23 years ago mobiles can kill' (25 July 1999). The campaign did not abate in the following year, when the newspaper suggested that the phones 'could make your brakes fail' (9 January 2000); that there may be a 'cancer link' (26 March 2000); and (perhaps more worryingly) that they could 'reduce your sex drive' (16 April 2000). However extravagant some of their claims, the *Sunday Mirror*'s campaign was organized around the compelling message that future scientific investigation would reveal the dangers of exposure. In an editorial, the newspaper asked, 'are mobiles the new tobacco?' (7 March 1999). The *Daily Mail* also featured the issue prominently (e.g. 'Loss of memory link', 1 March 1999; 'Harm to unborn babies', 24 April 1999), and frequently quoted the principal UK media experts on EMF damage, Roger Coghill and Dr Gerard Hyland.

When Cathy Moran moved from the *Sunday Mirror* to the *Express*, the latter also adopted a campaigning stance on mobile phone risk. Reflecting its more middle-of-the-road position, the *Express* was less sensationalist than the *Sunday Mirror*. The *Express* innovated new angles beyond the sometimes almost random claims of harm that characterized earlier stories. *Microwave News* noted that the articles in the *Express* were the first in the UK to link mobile phones to the health problems of specific individuals (*Microwave News*, November/December 1997). The *Express* became the most vehement daily newspaper critic of the mobile phone industry and the government's assurances about mobile phone safety. It increasingly focused on publicizing anti-mobile mast campaigns, rather than colourful claims about handsets. The *Sunday Mirror* had begun to report on the banning of masts from schools (5 December 1999), but did not systematically develop this aspect. The *Express* went much further on the environmental angle, publishing articles about people living near mobile phone masts who had developed cancer (28 September 1997). They reported ill effects on school pupils in Sunderland from a mast (2 March 2000), and on a ruling in Leeds that a mast had to be removed (8 March 2000), culminating in specific advice on 'how to fight [the] mobile threat' (11 March 2000). Alongside such stories, *Express* editorials urged a precautionary approach to base station sites (28 February 2000). The *Express* became the most important conduit for anti-mast campaigning, directing many of its subsequent enquiries to the influential campaigners on Kent County Council.

At the personal level, at least, this campaign was driven by a sincere conviction of possible harm from EMF. Moran came to believe strongly that mobiles and masts might be harmful.[9] Handling hundreds of public enquiries, she dedicated a considerable amount of her personal time to the issue, and directed hundreds of enquiries to other sources of support. She saw raising public awareness about the possible dangers of mobile phones and masts as her opportunity to do something 'more worthwhile'. At a wider level, the particularly prominent role played by the *Express* can be located in its distinctive circumstances during the late 1990s. Under the editorship of the radical feminist Rosie Boycott, the paper self-consciously sought to modernize and distance itself from its past. Taken over by Lord Hollick, then adviser to New Labour, the paper became known for being more 'on message' than any other; Boycott sought a 'successful new identity ... ditching the paper's traditional conservatism and trying to catch a wave of Blairite popularity'.[10] In particular, the paper appeared to try to relocate itself at the centre of the public safety issues that continue to be a dominant theme in 'risk-aware' contemporary Britain. In September 2000, for example, it railed against the lack of compensation for victims of human variant CJD.

Despite some 'silly season' stories about mobile phones, even the *Sunday Mirror*'s coverage was self-consciously precautionary rather than alarmist. In an editorial it pointed out that it was 'not scaremongering over the health risks and we are not against mobile phones, indeed they have revolutionised our lives for the better' (13 August 2000). It was the suggestion of an effect on children that was particularly prominent. 'Sickly pupils "recover" after leaving cell phone mast school' was a typical headline. The story repeated the claim of long-time anti-mast campaigner Debbie Collins that the removal of her daughter from a school sited close to a mobile phone mast had led to dramatic improvements in her health. Discounting the scepticism of 'even the experts', she claimed that, 'she's a different child now – it's all the proof I need to convince me there is a link between those wretched masts and the health of children'. Another blamed masts for a tumour. Although, 'At first they were concerned only about its impact on the value of the property', a Leeds woman now linked the mast to her husband's cancer. 'Nobody can say it isn't down to the mast,' she said (*Express*, 28 February 2000). Even more common than the 'you can't prove it's not the masts causing harm' claims, it is potential harm at the centre of concern. Parents campaigning against a mast erected near their children's school because of anxieties that it might damage their children in the future is perhaps the most typical story in the UK and in many other countries. Arguably, this reflects a wider contemporary social anxiety about the relationship to children and their futures (Roberts et al. 1995; Scott et al. 1998).

The tone of reporting in the tabloid newspapers was often not dissimilar to that of more 'quality' outlets. The pioneering roles of the *Sunday Times* and BBC television continued. In a major report at the end of 1998, the *Sunday Times* asked again: 'Are we being told the truth about mobile phones?' (20 December 1998). An edition of the BBC's *Panorama* on 24 May 1999 was as influential as *Watchdog Healthcheck* had been three years earlier, and became a common media reference point. The programme was organized around 'worrying new research on mobile phone safety'. Referring to the American industry 'whistleblower' George Carlo, the programme announced that 'one of the industry's own leading experts says the public must be told'.[11] Representatives from Nokia, Motorola and the NRPB were aggressively questioned and appeared defensive and uncommunicative. The *Panorama* broadcast was an important influence in raising the media profile of the EMF and health issue to new levels by June 1999. In the same week the Education Minister David Blunkett called for 'urgent investigation' of masts on schools.

Anti-mast campaigns became more prominent in the rest of the UK as base stations began to appear more prominently with the introduction of the digital mobile network.[12] At the beginning of 1999, local protests against masts were becoming relatively common and a staple feature of local and regional newspapers. Reports on protests against base stations became a mainstay of the local and regional press. Some random examples include:

- 'Local protests reported against siting of masts', *Courier and Advertiser* (Scotland) 25 January 1999.
- 'Village joy as mast plans scrapped', *Press and Journal*, 6 March 1999.
- 'Local mast "victories" are reported', *John O'Groats Journal*, 26 March 1999.
- 'Plan rejected amid health fears', *Congleton Chronicle*, 28 May 1999.
- 'Schools in Greater Manchester are in the middle of the growing nationwide health row over power masts for mobile phones', *Manchester Evening News*, 29 May 1999.

The local press has been an important source of publicity because it is at a very local level that anti-mast campaigns operate. Highly sympathetic reports on local campaigns have appeared in countless local papers. 'You mast be joking' runs a typical headline in one newspaper on a local residents' blockade of an attempt to erect a mast (*Reading Evening News*, 27 June 2000). Coverage has been quite sustained in many cases. The Kent-based *Gravesend Reporter*, for example, carried headline stories – including one entitled 'Mobile Warfare' – on 27 July, 10 August, 7 September and 26 October 2000. Local newspapers – usually reluctant to commit themselves to any type of campaigning activity unless they feel assured of victory – have publicly identified themselves with the issue. Another Kent-based paper, the *Gravesend Messenger*, has demanded that the local council oppose any further masts without full planning permission. Another, *This is Buckinghamshire*, has gone so far as to have a web-based anti-mast campaign.[13] Their highly publicized efforts do not concern one particular mast, but contest the process more generally. Besides drawing local attention to individual anti-mast campaigns, this extensive local newspaper coverage has provided considerable moral support for otherwise isolated campaigners. Jo and Tony refer to the ease with which they managed to get the local free paper to cover the issue, and this was evidently a useful source of support in sustaining enthusiasm.

If the government enquiry under William Stewart was expected to allay media concerns about mobile telephony, it was unsuccessful. Press campaigning and promotion of health concerns did not abate despite the extreme precautionary stance of the IEGMP report. The principal newspapers leading the promotion of concern continued to reveal problems. The *Sunday Mirror* demanded a reduction of emission levels in line with American standards (13 August 2000). The *Express* ran a front page declaring 'Kids in mobile phone alert' and at 'growing risk' after Disney dropped plans to market phones in the US (25 November 2000). The *Daily Mail* discovered possible 'nerve damage' (24 October 2000). Mobile phone health fears reappeared in the national press at the end of November 2000 following the government announcement that safety measures, which arose directly from the Stewart Enquiry, were soon to be implemented. Safety leaflets about children's usage were to be distributed at the point of sale of mobile phones. For the newspapers that have pursued a campaign to raise the profile of health concerns, the announcement provided an opportunity to raise concern again.

Arguably, the number of campaigning newspapers actually grew. *The Times* consistently engaged in the promotion of mobile phone health issues for the first time at the end of 2000. After publicizing yet another cancer-based American lawsuit against mobile phone companies, *The Times* announced that £5 billion had been wiped off Vodafone shares as a result (29 December 2000). The newspaper also publicized Mast Action UK's campaign on the same day. At the beginning of 2001, 'disturbing' studies continued to be revealed (*Daily Mail*, 16 January 2001) as well as reported tumours (*The Times*, 1 January 2001), and alleged litigation (*The Times*, 4 January 2001). New American research, which again disproved any connection between mobile usage and brain cancer, was barely reported in the British media (Muscat et al. 2000; Inskip et al. 2001). Mobile phones also figured prominently in media reports after the Stewart Report in relation to bullying and rising crime figures among children; mobile phone thefts were described as a 'playground plague' (*Daily Mail*, 17 October 2000) and a 'mobile menace' (*Evening Standard*, 10 July 2000). Overall, the British media has maintained a remarkable fascination with the 'phantom risk' of mobile phone emissions and even their connection to other social problems.

Media, public or government concern?

The campaigns against base stations have strengthened following the increased media exposure. The national media have played a direct role in cohering objections to the erection of masts. Following the promotion of Kent County Council Liberal Democrat group's 'Mast Action Newsletter' by the *Express* in March 2000, over 150 phone calls were received within 24 hours from individuals all over the UK.[14] Influential individuals, institutions and government began to orient themselves to media-driven concern about mobile EMF.

In response to the wave of press coverage in 1997 and 1998, some institutions and organizations began to express health concerns and even take specific precautionary measures. In August 1997, the National Union of Teachers consulted the Health and Safety Executive about possible health risks. In November 1998, Sir Richard Branson made his health concerns about mobiles public, recommending the use of safety devices to his employees (*Sunday Times*; *Sunday Business*, 15 November 1998). In a decision on 2 June 1999, the Metropolitan Police announced that officers were being told to limit mobile calls as a 'purely precautionary measure'. Specific reference was made in the press release to the *Panorama* programme of 24 May. The government also began to respond, most importantly by setting up the Independent Expert Group on Mobile Phones (IEGMP) in March 1999.

A further trigger to the heightened profile of campaigns against masts was the response of government and public authorities. At the end of 1998 and the beginning of 1999 local protests against the siting of radio base stations had first emerged in Scotland. Following a report by Friends of the Earth, indicating a more restrictive approach to mast siting by some local councils, others followed. By the end of March 1999, six Scottish councils had adopted a precautionary policy. The campaign then became more dynamic in the rest of the UK, given added momentum by the growing concerns about mobile phone handsets. The House of Commons Science and Technology Committee concluded an investigation of the issue in September 1999 with a call for massively increased research funds and an 80 per cent cut in permitted radiation levels from phones, which added to the speculation about health effects. Political interest in the issue continued to grow. In November 1999, Howard Stoate MP proposed an early day motion in the House of Commons which was signed by 181 MPs, calling for restrictions on siting of masts and funding for research into the possible health risks. At the local level, by the end of 1999, councils across the UK were experimenting with attempts to impose their own unilateral bans or restrictions on mobile masts, particularly those sited near to schools.

The *Sunday Mirror* boasted that after 18 months of campaigning against mobile phones they 'had blazed a trail all the way to the heart of government' (13 August 2000). Their claims were not without foundation. Certainly the most important moment of the mobile health story – the Stewart Enquiry and subsequent Report – was commissioned to respond to 'concern' rather than a concerted build up of scientific evidence. Announcing the formation of the Stewart Enquiry in 1999, the Minister for Public Health, Tessa Jowell, stated: 'In recent years research interest in the effects of mobile phones has increased. To date there has been no consistent evidence suggesting risk to health but there is continuing public concern about the possibility. It would be wrong to ignore that concern' (IEGMP 2000: 1). Specifically, the issue of children was singled out for reasons of 'public concern': 'In giving special attention to schools, the Expert Group was responding very largely to public concern rather than any proven health hazard' (IEGMP, 2000, Clarification of Issues Discussed in the Report).

If public concern was far from overwhelming, the Enquiry was initiated in anticipation of such a development. Media coverage was apparently here taken as an early warning.

Explaining her proactive approach before a House of Commons Select Committee, Jowell highlighted the 'considerable amount of media interest, media concern about the potential ill effects on health from mobile phones'. She went on to make clear that in, 'an area like this ... it is very important that we ... work very hard to keep ahead of public *anxiety* ...' (House of Commons 1999: 1; emphasis added). It was in this anticipatory and precautionary spirit that the IEGMP was commissioned in March 1999. In this useful formulation, the media are 'ahead' of (anticipated) public anxiety.

At the same time, more traditional democratic pressure undoubtedly played an important role in driving government to an 'anticipatory' response. In so far as any single issue dominated MPs' postbags during 1999, it was from constituents angry at the siting of mobile masts. The Stewart Report detailed the number of complaints to politicians and the state: 600 letters to ministers – mainly concerning masts; 85 letters to MPs replied to by health ministers; 80 letters to the Department of Health; 350 to the DETR on planning and environmental issues; and 157 to the Department for Education and Employment about the location of base stations in or near schools. They went on to note that the mast issue was the most popular early day motion currently tabled in the House at the time of writing (IEGMP: 2). In these terms the government's action would appear to be a good example of 'responsive' and 'open' government. However, it should be remembered that the relationship between this more conventional public pressure and a major government-sponsored enquiry that embraced the spirit of precaution from the outset is far from direct. In their own terms these letters of complaint principally concern issues of planning, location and environment – not health concerns about electromagnetism alone. Judging by letters of mast complaint I have seen, health would undoubtedly have been mentioned as an important factor – but only among the other issues of 'planning, location and environment'. These health concerns – as has been established – were far from spontaneous. It is only through interaction with the long-running and intensive media health problematization of mobiles that this factor came to prominence. In other words, more apparently conventional democratic public pressure cannot be separated from the highly effective media campaign.

While the media has not ended Britain's 'love affair' with the mobile, they have perhaps made it a more anxious relationship than it might otherwise have been. At least to some extent they have a compelled a significant government response 'anticipating' public concern.

The 'old' media certainly appear to have had some direct effect on government policy related to the 'new' media – separately from any real impact on public consciousness. In the 'risk society' it may well be that the media have become far more than just the messenger, but a substitute for an increasingly diffuse and intangible 'public concern'. Whilst it is unlikely that media-generated anxiety about mobile phones in the UK will ever again reach the peaks of 1997–99, it is equally unlikely that we have seen the last of such episodes. Precaution has become a political imperative in the 'risk society', ensuring an important audience for even the most marginal claims of potential damage to our health and children.

Notes

1 The IEGMP chaired by Sir William Stewart was commissioned by the Department of Health in 1999 to investigate scientific knowledge of mobile risk. Its conclusions were published in May 2000.
2 It is instructive to note that it was the 'possible threat' to children which constituted the core reaction as reflected in the national newspaper coverage on the day of the release of the final report (11 May 2000). The Report's conclusion that there was no evidence of harm was virtually ignored. Newspapers also rightly pointed to the curious position of the Report's 'advice' that it was a personal choice whether parents allowed their children to use them.
3 The global mobile industry's principal forum on health risks, the annual IBC 'Is There a Health Risk?' conference had a record international representation of 38 nations at its December 2000 London event.
4 After the DETR announced new limits on mast erection, in April 2001 the Commons Trade and Industry Committee demanded further restrictions, for example.
5 Interviews were carried out with some 20 campaigners, as well as countless shorter exchanges with prominent individuals in the field internationally during the summer of 2000.
6 Margaret Dean's witness statement to the IEGMP.
7 The *Northern Echo* 'revealed' in 1995 that some mobiles emitted radiation far above EC safety guidelines. The focus for the story, which continued into 1996 (18 and 19 June), was lobbying by Durham MEP Stephen Hughes who urged safer EC limits and made the paper aware of his actions.

8. On 14 January 2001 Leake published a report on research linking mobiles to eye cancer and in the process ignoring American studies which found no negative effects (14 January 20001). He did not mention the important American studies of Muscat et al. (2000) and Inskip et al. (2001) which were major news items in the US.
9. Interview on 23 August 2000.
10. Donald Trelford, *Evening Standard*, 25 January 2001.
11. George Carlo was previously the scientific head of the American mobile industry's research programme on health issues. Carlo is now the leading voice of alarm in the US, widely promoting his book (Carlo and Schram 2001).
12. Despite its better service, digital phones require a far more extensive network of masts.
13. thisisbuckinghamshire.co.uk/buckinghamshire/Bucks_Matters/mobile_phones/index.html
14. Mobile Phone Mast Action Letter No.3 (2000), Liberal Democrat Group, Kent County Council, Maidstone.

References

Axford, B. and R. Huggins (2000) *New Media and Politics* (London: Sage).
Beck, U. (1992) *Risk Society: Towards a New Modernity* (London: Sage).
Brauner, C. (1996) *Electrosmog: A Phantom Risk* (Zurich: Swiss Reinsurance Co.).
Carlo, G. and M. Schram (2001) *Cell Phones: Invisible Hazards in a Wireless Age* (New York: Carroll and Graf).
Croteau, D. and W. Hoynes (1999) *Media/Society* (London: Sage, 4th edn.).
Diani, M. and P. Donati (1999) Organisational Change in Western European Environmental Groups: A Framework for Analysis, *Environmental Politics*, 8(1): 13–34.
Donati, P. (1997) Environmentalism, Postmaterialism and Anxiety, *Arena Journal*, 8: 147–172.
Foster, K. R. et al. (1999) *Phantom Risk: Scientific Inference and the Law* (Cambridge, MA: MIT Press).
Giddens, A. (1990) *The Consequences of Modernity* (Cambridge: Polity Press/ Stanford, CA: Stanford University Press).
Habermas, J. (1989) *The Structural Transformation of the Public Sphere* (Cambridge, MA: MIT Press).
House of Commons Science and Technology Committee (1999) *Minutes of Evidence* (London: Stationery Office).
Hyland, G. (2000) Physics and Biology of Mobile Telephony, *Lancet*, 356: 1833–1836.
Independent Expert Group on Mobile Phones (2000) *Final Report: Mobile Phones and Health* (Didcot: National Radiation Protection Board). Available at url: www.iegmp.org.uk.
Inskip, P. D. (2001) Cellular Telephone Use and Brain Tumours, *New England Journal of Medicine*, 344(2): 459–519.
Melucci, A. (1989) *Nomads of the Present* (London: Hutchinson).
McQuail, D. (2000) *McQuail's Mass Communication Theory* (London: Sage, 4th edn.).

Muscat, J. E. et al. (2000) Handheld Cellular Phone Use and the Risk of Brain Cancer, *Journal of the American Medical Association*, 284(23): 3001–3007.

Roberts, H. et al. (1995) *Children at Risk? Safety as a Social Value* (Buckingham: Open University Press).

Scott, S. et al. (1998) Swings and Roundabouts: Risk Anxiety and the Everyday Worlds of Children, *Sociology*, 32(4): 689–706.

World Health Organization (2000) Fact Sheet No. 193: *Electromagnetic Fields and Health* (Geneva: WHO).

Part IV
Problems of Risk Communication

12
The Rhetoric of Risk and Responsibility: Understanding the German Public Debate on EU Enlargement

Matthias Ecker-Ehrhardt

Introduction

Like most contributions to the present volume, this chapter is not about 'objective' risks but about ways of 'talking politics' in terms of a more or less explicit language of risk. Unlike the common focus on technological risks, I will discuss 'risks' as social phenomena in a broader sense, as perceptions of political or socio-economic consequences produced by political action. These enter political debates and become common and legitimate imperatives for certain political choices due to their quality as 'good reasons' *vis-à-vis* more or less diffuse but seriously harmful (side-) effects of political (in-)action, such as massive unemployment, political disintegration or even war. In this way, and beyond the given focus of 'risk research' as the common denominator of this volume, I argue that we have to look *beyond norms and identity* if we want to come to a comprehensive understanding of how 'ideas' shape political conflicts over international issues. Following a rather meta-theoretically driven endeavour to define a non-rationalistic research programme, many 'constructivists' have become obsessed with proving the causal and constitutive effects of norms and identity,[1] ignoring the fact that definitions of empirical problems and cause-and-effect relationships are essential for political actors to come to terms with policy choices and critical questions from their domestic constituencies.[2]

In this way, the European Union's eastern enlargement is an interesting case in point. The way in which member states were able to agree on the full accession of several Central and Eastern European countries (CEEC) has attracted a lot of scholarly attention. For most authors the definition of normative appropriateness played a key role in inducing Western leaders to open negotiations in 1998 (cf. Schimmelfennig and Sedelmeier 2002), finally leading to the accession of eight applicants from Central and Eastern Europe in 2004. While this seems true, the common practice of referring to norms of a 'uniting Europe' and community-building around common values alone seems partial at best. How was it possible for Western European governments to generate a common understanding of completely new and unexpected circumstances after 1989? How were they able to build domestic support on such a risky endeavour as the full integration of a good number of former communist countries? As will be argued, the definition of normative duties gained ground because of converging expectations of a risky future without a proactive EU accession and enlargement policy based on common interpretations of empirical parameters like economic disaster and its negative effects on democracy and security. Later, this consensus was modified in part by blending current lines of reasoning with alternative expectations of 'unmanageability' and 'pauperization' to justify conditions of an accession that turned the early 'Yes' on enlargement into a restrictive and cautious 'Yes, but …'. Hence an analysis of the early debate on EU enlargement broadens our understanding of political deliberations in which a rhetoric of risk complements normative 'responsibilities' for political action in terms of national interests, but also in terms of an altruistic calculus of 'helping' and 'solidarity'.

In order to get a grip on how this worked empirically in the case presented here, I focus on parliamentary speeches and newspaper editorials in Germany on the enlargement issue, i.e. a corpus of texts that constitutes an important part of the public debate there (cf. Ecker-Ehrhardt 2002, 2004). My argument draws on a thorough content analysis of German newspaper editorials which appeared in the influential *Frankfurter Allgemeine Zeitung* (FAZ), *Süddeutsche Zeitung* (SZ) and *die tageszeitung* (taz) (n = 232), as well as of parliamentary speeches held in the German Bundestag (n = 256) with reference to the enlargement issue, from the beginning of the debate in autumn 1989 until the German federal elections in 1998. By promoting or criticizing certain courses of action, participants in the debate were trying to give 'good reasons' as parts of more or less complex 'arguments' (Habermas 1981; Toulmin 1996). Restrictively defined as an attempt to justify a preference on specific aspects of the enlargement process or enlargement in general, such 'enlargement preferences' functioned as the respective units of text analysis.[3] The aim of this analysis was to establish which attributes of arguments are articulated by certain authors, or by them more often than by others, indicating a latent discursive structuring of interpretations. Following the common sociological understanding of alternative kinds of ideas, I broadly distinguish between *analytical and normative ideas* as important attributes of arguments. While analytical ideas denote social beliefs concerning how the world 'is', i.e. knowledge about facts and causal relationships, *normative ideas* denote social beliefs regarding how the world 'ought to be', i.e. in terms of values, general norms of appropriate behaviour, and duties given for specific identities of single actors or collectives.[4]

How 'responsibilities' were defined by using norms and narratives

Long before Western societies were even speculating about 'velvet revolutions' emerging out of a seemingly stable hegemonic system behind the 'Iron Curtain', Polish intellectuals were deliberating over Poland's and Germany's role in a free Europe, together on their way toward a common European Union (Kerski 2001). After coming to power in 1989, the new Polish elites proved to have a very clear vision of what a 'new Eastern policy' embraced by Germany and the European institutions should look like: after a transition period – and Western aid – Poland, like others, would be able to become a member of a demanding and ambitious project like the European Community. While some analysts diagnosed a 'cognitive gap' and even 'mental confusion' among Western leaders confronted with these demands (Sedelmeier and Wallace 1996), this diagnosis needs specification. First, it is fair to say that many observers were able to anticipate them quite early on. With respect to the German domestic debate, prominent actors from the political centre and civil society not only anticipated these ambitions as early as autumn 1989, even before they emerged in interviews available to the German public. Indeed, in light of the Community's long enlargement tradition and Germany's specific role as a promoter of European integration (Gower 1999), many participants in the German debate were convinced from its very beginning that the European framework would be one of the few feasible instruments for handling problems and uncertainties caused by the ending of the Cold War and Germany's new role as a full sovereign state 'on probation' (i.e. under the suspicious eyes of its partners). What is puzzling about this debate, though, is less how policy alternatives were defined *per se* (cf. Ecker-Ehrhardt 2004) and more how specific policy alternatives were chosen in the wake of a remarkable convergence of expectations over the risks and opportunities of the political, social and economic transformation of the CEEC.

A singularly striking feature for most was the fact that throughout Europe the whole enlargement debate was permeated from the start with narratives of normative duties and obligations towards the applicant societies (Sedelmeier 2000). Germany was seen to be in a special position due to its historical background, and normatively obliged to act as an 'advocate' (*Anwalt*) of the applicant states *vis-à-vis* European institutions and member governments – a role repeatedly acknowledged by German officials and formally declared in a series of bilateral agreements endorsing 'good neighbourhood', 'friendly co-operation' or 'partnership' in 1991 and 1992, respectively.[5] The responsibility for war, occupation and postwar 'exclusion from freedom and wealth' (Yalta) were not the only 'historical debts' to be settled by proactive German engagement at the European level, as Germany 'would not have reached [its] major national goal, reunification, without the desire of our Eastern neighbors and friends for freedom'.[6] In particular, Polish, Czech and Hungarian merits in bringing down Soviet hegemony were therefore advanced as important arguments for meeting common standards of reciprocity and showing 'gratitude' by 'doing good'.

Last but not least, in terms of 'identity politics' the definition of a shared European identity, a common history and cultural heritage paved the way for definitions of a comprehensive political collective and integration as a common good in itself. In this way, metaphors of Europe – including member states and the Central and Eastern European applicants – as a 'house', 'family' or even 'body' injected a strong spirit of 'essentialism' into the debate (cf. Hülsse 2003). Some participants went so far as to declare enlargement to be a 'reunification of the continent',[7] ending an 'unnatural separation'[8] of Europe. In consequence, Germany's 'poor cousins' – the most extensively used expression throughout the German debate on enlargement – were seen as entitled to expect strong support for 'their return home to Europe'.[9]

Hence Germany was in a state of 'moral entrapment' from a very early stage: Virtually no participant in the debate denied any already proclaimed political responsibilities, since that would have caused a serious loss of credibility and reputation, a kind of shaming by political opponents. In this way the German debate preceded and surpassed the debate at the European level in intensity and variety of moral arguments (cf. Friis 1998; Schimmelfennig 2003). Nevertheless, even if direct counter-arguments may come close to an act of political suicide at a given stage of a morally driven discursive closure, political debates in general do not necessarily end with such consensus on 'responsibilities' of the linguistic foreground. 'Responsibilities' rest on some diagnosis of empirical parameters that can be contested, opening up avenues for renegotiations. Moreover, to argue for or against a certain policy alternative in the public arena requires a convincing link between 'responsibilities' for the welfare of others as policy goals on the one hand, and strategic choices on the other. As has been convincingly argued by Axelrod (1976), Hall (1993) and Sabatier (with Jenkins-Smith, 1999), participants taking part in political deliberations are permanently in need of credible assessments of the consequences that certain policies may entail, supplied through experiential or scientific knowledge of cause-and-effect relationships and packaged in complex 'paradigms'.[10] The existence of such paradigms is necessary to cast policy alternatives in the shape of powerful 'duties' in the light of accepted 'responsibilities'. In the case of enlargement, this complex interplay of norms, narrative, parameters and paradigms is crucial for an understanding of how the late consensus on a conditional 'Yes, but ...' for enlargement was finally reached and brought into the European process of negotiations by the German government.

What is at stake: building problems out of parameters and paradigms

To start with, a common understanding of a serious problem has to be established before defined responsibilities can be translated into duties to solve it. In the case of the German debate on European enlargement this was achieved through a narrative of the devastating consequences of 'forty years of communism', i.e. of authoritarian rule and planned economy:

> [T]here is no developed infrastructure, no functioning bureaucracy, no sufficient communication networks and educational facilities, instead there are, with respect to market conditions, a highly distorted economic structure, an uncompetitive heavy industry, massive pollution, and an extensive but unproductive agricultural sector, compared to Western standards.[11]

What made this diagnosis a basis for risk perceptions were widely held beliefs on how prosperity, social cohesion, democratic stability and peace are causally connected. Based on a well-established and taken-for-granted paradigm of the *economic fundaments of democratic institutions*,[12] prospects for the 'young democracies' in the CEEC were rated as poor. Given their economically disastrous situation, with high unemployment, it was deemed inevitable that social deprivation would deepen social and political conflicts, preventing any consensus for a new libertarian polity to evolve. Moreover, as the Germans have learned as one of their main 'historical lessons', social deprivation fosters anti-democratic movements from both extremes of the political spectrum, leading to domestic and even international conflict.[13] As Wolfgang Koydl, among others, pointed out in the *Süddeutsche Zeitung*:

> [S]tability in Central and Eastern Europe is in danger. Not due to any external threat, i.e. by the Soviet Union, but from inside: permanently neglected national and ethnic tensions, social conflict, the pauperisation of broader strata of society. All this sows the seeds for civil war and border disputes, opening the doors for ruthless populists and even fascist regimes.[14]

Moreover, pressures to stabilize the Eastern neighbours emerged out of the common interpretation of Europe as a region of complex *interdependence* with Germany in a precarious 'central position' (*Mittellage*), i.e. exceptionally vulnerable to instabilities and their effects: 'we have a special interest [in unifying Europe]. Our country lies in the center of the continent. We have the most neighbours and the longest borders' (Helmut Kohl).[15] Deliberations on the possible failure of Central and Eastern European transformation thus immediately gave way to perceptions of German risks involved in the debate over enlargement.

What is effective: choosing policy alternatives to solve problems

If they are to build sound arguments for either transnational 'responsibility' or 'national interests', speakers have to make it clear what political measures they are advocating to realize their respective goals, i.e. how their own policy relates to desirable outcomes *vis-à-vis* given assumptions on cause-and-effect relationships embodied in 'paradigms'. As argued, the early consensus in favour of enlargement rested on a common definition of the poor (economic) situation as well as on complementary paradigms such as the economic conditions of democracy and European interdependence which made the transition of the CEEC a desirable but risky endeavour for Western Europe.

First, eastward enlargement was seen by many observers as a panacea for major problems at stake. Following the dominant macroeconomic thinking on improving the *'terms of trade'* (Baldwin and Venables

1995), the removal of all trade impediments, it was argued, would encourage competition on both sides of the former Iron Curtain, inevitably leading to economic restructuring and a significant increase of productivity and international competitiveness (Heitzer-Suša 2001: 57–60). Subsequently, and in accordance with a paradigm on the *economic fundaments of democratic institutions*, gains in prosperity would ease distribution problems and facilitate the political transition to a stable democratic system. It is important to note that these expectations were common during the first years, and could even be heard outside the liberal and conservative parties and certain newspapers well known for their liberal economic discourse, like the *Frankfurter Allgemeine Zeitung*.[16]

Moreover, Germany's own historical experiences commend *multilateralism* and European political integration as an effective (and normatively) 'appropriate' means to stabilize the transition from authoritarian rule to democracy (cf. Baumann 2002); or, as Chancellor Kohl repeatedly stated, 'the only effective assurance against nationalism, against power politics, and against war'.[17] In functional terms, such arguments can be understood as an attempt to 'lock' democratic institutions into strong European institutions and a strategy of 'stabilizing the domestic political status quo against undemocratic threat' (Moravcsik 2000: 220). With respect to the German domestic debate during the 1990s, such arguments made up a good part of the 'linguistic foreground', produced by and resonating with corresponding beliefs of causal relationships between enlargement policy and the given goal of 'stabilizing' the applicant states and European peace in general. As the German Foreign Minister Klaus Kinkel euphorically stated:

> The founding fathers of the European Community had a great vision after the war: to bring enduring peace to our continent, which was worn down after centuries of fratricidal war. Growing integration has made war unthinkable in Western Europe and laid a fundament for prosperity never known before. ... The end of the conflict between East and West has given us the chance to extend this model to the whole of Europe.[18]

Last but not least, integrative *multilateralism* was seen as a precondition for Germany acting proactively at all, as any unilateral step on the international scene tends to evoke old reflexes in its European partners. In the face of this 'German predicament' (Markovits and Reich 1997), a

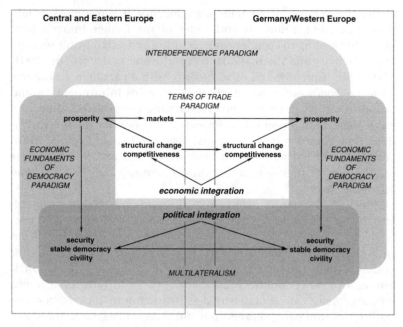

Figure 12.1 Paradigm underlying the 'Yes' of the German debate on enlargement

retraction from its earlier position on enlargement was seen as highly problematic (see e.g. Joschka Fischer, *Deutscher Bundestag* 13/241: 22199).

As can be seen in Figure 12.1, those paradigms of 'terms of trade', 'multilateralism', 'interdependence' and 'economic conditions of democracy' were complementary, functioning as parts of a comprehensive discursive 'map'[19] that enabled actors to converge on how enlargement related to the interests of the societies involved. Still, only in combination could 'responsibilities' and paradigms be used to formulate an ethical duty for Germany to act as an 'advocate' of enlargement at the European level. In sum, the early consensus over enlargement grew out of a thick cultural turf of widely held and complementary paradigms, norms and narratives. While narratives helped to define 'historical debts', paradigms were even essential in making economic and political integration an effective means of a prudent German foreign policy in the eyes of the debating elite.

Rhetorical backfire: from the early 'Yes' to the late 'Yes, but ...'

Intended and unintended outcomes of rhetorical strategies are what make of political debates emerging 'social facts' rather than pure aggregations of individual acts. What makes the enlargement debate an instructive case in this regard is the dynamic that developed after the early consensus on integration of the CEEC into the European Union was reached. Content analysis reveals how the early 'Yes' for enlarging the European institutions eastward gradually turned into a conditional 'Yes, but ...' over the course of the 1990s. During this process (1991–92), economic stability and prosperity rapidly became the major accession criteria for most of the speakers. Thereafter the question of 'widening versus deepening' (cf. Tewes, 1998) ended in a broad consensus on necessary reforms. Moreover, transition periods for farm subsidies and the free movement of labour became an issue and were finally successfully negotiated by the Danish presidency, accommodating pressures from the German and French governments.

How was this possible? Since normative responsibilities for the CEEC were undisputed, they must be seen as important, and perhaps driving, but more or less constant factors of the German enlargement debate. Therefore, they cannot account for the change of preferences over enlargement policy. One has to follow those rhetorical endeavours that were emphasized as a means of legitimizing the integration of the CEEC into the EU to reach a comprehensive understanding of how the final mixture of 'Yes' and 'But' emerged. This rhetoric backfired to a remarkable extent as it fuelled critical assessments of an 'unconditional enlargement'. In a nutshell, critics from various margins of the political spectrum soon began to question whether Western Europe really should open its borders to such an unstable region and potential source of insecurity. Important points put forward by such critics made their way into the broad consensus on enlargement by adopting and adapting widely held beliefs about the situation and cause-and-effect relationships.

To start with, counter-arguments soon began to gain ground in the face of an unexpectedly protracted economic struggle in the new *Länder* of the former GDR. Beginning with well-known spokesmen from the left of the German political spectrum such as the Party of Democratic Socialism (PDS), later even the Social Democrats (SPD) and the Greens started to articulate fears of *pauperization*: economic restructuring, they argued, would probably increase productivity, but at the cost of exploding rates of unemployment and poverty. Ultimately, 'neoliberal shock therapy' would lead to social deprivation and nationalism, the chances for stable democracy and peace in Europe would diminish rapidly, and the 'responsibility' assumed on behalf of the applicant societies would be poorly met (cf. Randzio-Plath, *die tageszeitung*, 23 October 1991). While enlargement was widely seen as a 'chance to gratify and stabilize young democracies',[20] it soon became a threat to the transition process, put at risk by 'hasty enlargement' (cf. Oldag SZ, 4 October 1995). What is striking about this argument is a common line of reasoning with regard to the economic requirements needed for democratic institutions to find support by their constituency beyond elite circles (Figure 12.2). In fact, both sides of the debate – the side that advocated rapid integration into the internal market and the side that opposed it – made use of the same paradigm, although each fed it with contradictory expectations of prosperity and poverty. In this way, respective counter-arguments became hard to negate as they not only rested on the same diagnosis of a catastrophic economic situation, but even shared parts of the analytical knowledge persuasively linking policies with normatively given goals.

Moreover, 'interdependence,' an already prominent way of defining causal linkages, made all these risks of an 'untimely opening up of borders' a threat to German interests. While liberals were claiming a common interest in economic restructuring and international competitiveness, proponents of a pauperization paradigm saw Germany threatened by rising rates of unemployment leading to social deprivation and even, for some observers, a rise of the extreme right. Those expectations were fuelled in

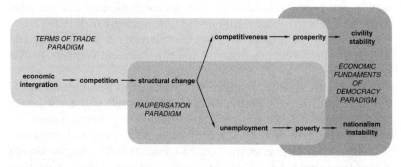

Figure 12.2 Alternative paths to the paradigm on the economic conditions of democracy

part by a complementary interpretation of social deprivation as a main cause of migration (Heitzer-Suša 2001: 62–4). This, it was argued, would aggravate unemployment in two ways: economic restructuring and migrants from the CEEC 'flooding' the German labour market.

Finally, the enlargement issue was articulated with a Eurosceptical discourse, identifying European integration with negotiations *ad infinitum*, bureaucratic inaction and Germany's role as a 'net contributor' (*Nettozahler*). Having 'learned the lesson' of southern enlargement, European institutions would soon become 'unmanageable' due to more 'poor' at the table negotiating for redistributive measures on their behalf and empowered by consensus-oriented decision-making (cf. Oldag SZ, 4 October 1995). Important addenda to the 'unmanageability' argument were the anticipated costs of enlargement. As early hopes of a rapid recovery of the East German economy were disappointed, old demands for a reduction of Germany's net contributions to the EU budget gained momentum, and the new costs were regarded as one of the most serious 'German' risks from enlargement (cf. Waigel, *Deutscher Bundestag* 13/247: 22206–7).

In the end, an argumentative coalition pressuring for a restrictive stance on the enlargement issue prevailed. While accepted as a normative duty, the consensus on the desirability of enlargement rested on a large set of conditions that the applicant states had to meet before they would be allowed to become full members. Moreover, when the Schroeder–Fischer government came to power in 1998, free movement of labour became the most crucial point of concern for the new Chancellor (cf. Schroeder, *Deutscher Bundestag* 13/247: 23068 and 14/3: 66). In 2002 the accession negotiations ended when all participants agreed on a transitional period of seven years, taming German domestic critics to a certain extent. With respect to risk perceptions governing this shift, underlying paradigms can help to understand how this was accomplished (Figure 12.3). Establishing an alternative reading of how economic processes relate to the interests of European societies caused an integral part of the 'cognitive map' of the German debate on enlargement to be replaced. With analytical paradigms as essential 'switchmasters' (Weber) at work, most of the parameters changed into their opposite – from prosperity to poverty, indeed from stable democracy and peace to instability and conflict.

Instructively, the respective shift toward a 'Yes, but ...' was not simply a change in the relevance or size of a 'Yes' and a 'But ...' in terms of competing 'camps' in the public sphere, it was a change in the dominant mode of debate over enlargement. In this process *the hegemonic consensus*

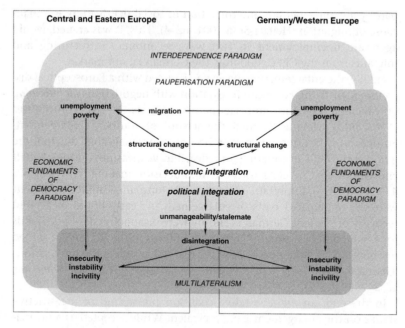

Figure 12.3 Paradigm underlying the 'But' of the German debate on enlargement

among all major parties and commentators from the elite media was modified by integrating *risks and conditions into their arguments on enlargement.* Using internal reforms as an example, quantitative analysis of identified arguments can be used to illustrate the increasing association of statements in favour of enlargement with risk perceptions and conditions on the level of complete texts (speech, editorial). As can be seen from the results presented in Figures 12.4 and 12.5, statements in favour of enlargement were initially unaccompanied by any reference to risk of 'unmanageability' and necessary internal reforms in most of the texts. This low association changed in subsequent years and after critical observers from the media had established their views of risks and pressured for institutional reforms as an 'inevitable preparation' for enlargement. Later, after perceiving the costs of enlargement, policy reforms and an adaptation of the EU finance system became a common condition in the context of arguments for enlargement. In this way, the 'Yes, but ...' seems an appropriate label not only for the debate as a phenomenon at the macro level, but for individual argumentative practice at the macro level as well. Hence critics succeeded in keeping up a comprehensive

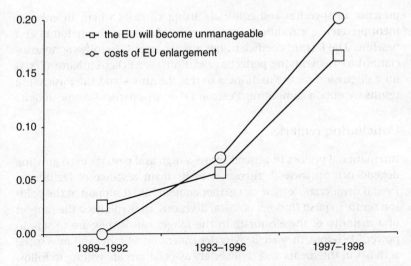

Figure 12.4 Argumentative integration of the risks of enlargement

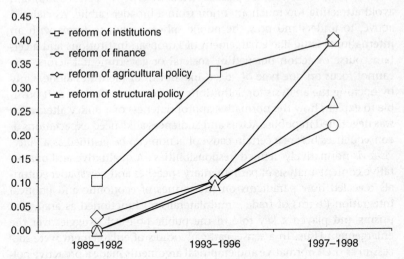

Figure 12.5 Argumentative integration of the conditions for enlargement

consensus in the domestic arena, viable and specific enough, though, to commit the German government to taking the corresponding stance at the intergovernmental level (cf. Lippert 2002).

Argumentative integration is measured here in terms of an association of arguments with an enlargement preference. Entries are Jaccard coefficients (cf. Batagelj and Bren 1993), measuring the proportion of mutual presence in speeches and editorials, using all cases where an enlargement preference, conditions or risks under observation were found as a baseline. The Jaccard coefficient has a range from 0 (signifying no association) to 1 (signifying perfect association). See Ecker-Ehrhardt (2004) for a discussion on the usefulness of this measure – and the misleading results we obtain computing 'Pearson's r' on this particular type of data.

Concluding remarks

International politics in general shows a high and perhaps even growing dependence on societal discourses as its main resource of legitimacy. Even if democratic leaders can gather enough initial support at the political centre to push through political decisions, they still need the support of a majority of the electorate in the longer run if they are to stay in power. Therefore, they are heavily dependent on whether and how social activists in the media and civil society associations are willing to follow their reasoning on specific issues if they are to keep criticism within the narrow margins of 'routine politics' (Habermas 1992; Peters 1993) and avoid attracting too much attention from a broader public. As claimed above, to understand how the public sphere comes to terms with an international issue like enlargement of European institutions and a certain course of action pursued by societal or governmental actors, one cannot focus on one type of 'ideas' like norms or identities alone. Only by opening the analysis for definitions of 'facts and effects' was it possible to explain how the normative appropriateness of a policy alternative was negotiated in public. Actors and audiences alike need expectations of consequences before a certain cause of action can be justified as a 'duty' vis-à-vis normatively defined 'responsibilities'. A qualitative and quantitative content analysis of parliamentary speeches and newspaper editorials revealed how paradigms on the virtues of economic and political integration ('terms of trade', 'multilateralism') functioned as analytical prisms and played a key role in the public political struggle over the enlargement issue. In a sense, early advocates of enlargement were successful as their normative and empirical arguments made a proactive policy toward the Central and Eastern European countries into a duty. Nevertheless, based on a narrative of economic disaster and political instability in the CEEC, the rhetoric 'backfired' to a remarkable extent. In the light of already circulating paradigms on the 'economic conditions for democracy' and 'interdependence', those interpretations fostered the definition of risks of enlargement, which was used as a justification for delaying full integration. As a consequence, some activists' arguments in this particular game concerning enlargement failed when their own constructions of risks – which were intended to function as 'good reasons' for enlargement – backfired, becoming justifications for delaying the full integration of Central and Eastern European societies into the EU.

With respect to research on international relations in general, this seems important even beyond a foreign policy analysis framework conducted in a constructivist spirit. Due to a 'mediatization' (Kepplinger 2002; Schulz 2004) of politics in general, an emerging sphere of 'global governance' has been confronted with an intensification of opinion formation and assertiveness among the wider public. As Zürn (2004) has convincingly argued, 'executive multilateralism' seems to be under pressure and becoming 'reflective', i.e. it appears to be seen not only in terms of a growing critical public awareness of international institutions like the World Bank or the World Trade Organization (Rucht 2002), but also in terms of a withdrawal of 'permissive consensus' towards European integration at the regional level. In all these cases, the normative grid of relevant responsibilities and recognized authorities is obviously a new topic of domestic debate, making those institutions the targets of conflicting expectations by different sectors of society. As in the case of European integration, the build-up of authority rests on accumulating experiences from prudent policies and coping successfully with serious problems. Consequently, a thorough analysis of the main paradigms and rhetorical strategies underlying political debates on the risks and virtues of 'global governance' seems indispensable to understanding how 'authority' beyond the nation-state is negotiated. Moreover, the scope of analysis must not remain confined to technological risks but must open its focus to all expectations of serious but potential harm, which permeate public discourse. Their relevance, one might speculate, will continue to grow as a primary resource of actors to be heard in a highly competitive race for public attention.

Notes

1. Most prominent proponents, like Wendt (1999), Checkel (1998), Adler (1997) and Finnemore and Sikkink (1998), to name only a few, have even tried to identify the entire constructivist research programme with a practice of proving the crucial role of norms and identities for social processes in general.
2. This is not to say that these aspects have not played a prominent role in some branch of the scientific debate; to the contrary, some research has been launched on 'cognitive maps' (Axelrod 1976; Shapiro et al. 1988), and even more on 'epistemic communities' (Haas 1992), as will be argued below. Nevertheless, the

integration of those efforts into a broader perspective on domestic debates on international issues is still lacking.

3   Technically, the result of this discourse analysis has been a matrix that identifies arguments (defining the 'rows' of the data set) with its possible components as variables (coded present or not present), such as preferences, norms, collectives, consequences, or historical narratives. The argumentative practice of 'giving justifications for policy alternatives' is the basic information stored by this matrix, making possible a subsequent quantitative analysis of modal types of combinations prominent in the debate.

4   What makes them different as social phenomena are essentially alternative ways of how infringements are handled (cf. Luhmann 1986). Analytical ideas can archetypically be proved 'wrong' by contradictory evidence and are thus the classic object of experiential learning. Normative ideas are not falsifiable – at least in the short run – by deviant behaviour; but behaviour itself that is deemed 'wrong' or 'inappropriate'. In this way it seems best to understand both aspects more as qualities of empirical ideas than as clear-cut categorizations of them. Both 'falsification' and 'obligation' can apply to different ways in which actors cope with deviance in empirical circumstances.

5   The bilateral treaties with Bulgaria and Romania affirm German 'support' for starting negotiations over an association soon; passages to be found in those treaties signed with Poland, Hungary, Slovakia and the Czech Republic go even further and affirm German support in 'bring[ing] about conditions for full integration' into the European Communities (see Ecker-Ehrhardt 2004 for an overview).

6   'Ohne den Freiheitswillen unserer östlichen Nachbarn und Freunde hätten wir unser wichtigstes nationales Ziel, die Wiedervereinigung, nicht erreicht' (Klaus Kinkel, in *Deutscher Bundestag* 13/224: 20431).

7   Nonnenmacher, *FAZ*, 19 June 1996.

8   Middel, *Die Welt*, 31 March 1998.

9   Helmut Kohl, citing Vaclav Havel, *Deutscher Bundestag* 13/181: 16223.

10 By paradigm I mean 'framework of ideas and standards that specifies not only the goals of policy and the kind of instruments that can be used to attain them, but also the very nature of the problems they are meant to be addressing' (Hall, 1993: 279). Following Hall, I prefer to call them 'paradigms' rather than 'implicit theories' (assuming consistency and falsifiability; cf. Hofmann 1993) or psychological 'cogmaps' (which were conceptualized on an individualistic model by Axelrod (1976), although Shapiro et al. (1988) are an instructive exception); see Hall (1993: 280) for an instructive discussion.

11 'In Polen gibt es keine entwickelte Infrastruktur, keine funktionierenden Verwaltungen, keine ausreichenden Kommunikationsnetze und Bildungsmöglichkeiten, statt dessen ein, an Marktanforderungen gemessen, noch völlig verzerrtes Wirtschaftsgefüge, eine konkurrenzunfähige Schwerindustrie, eine massive Umweltverschmutzung und eine zwar umfangreiche, aber nach westlichen Maßstäben wenig produktive Landwirtschaft' (Gerd Poppe, *Deutscher Bundestag* 12/50: 4086).

12 For the scientific version of this paradigm, see Lipset (1959), Dahl (1989), Burkhart and Lewis-Beck (1994).

13 In this way, a German variant of the 'democratic peace paradigm' could be detected in many arguments, but always with reference to revisionist

movements, never in the way of 'democratic peace' as explained by liberals along the Kantian line of thinking (see Moravcsik 1996 as a prominent example).

14 'Ohnehin ist die Stabilität Mittel- und Osteuropas gefährdet. Nicht durch eine äußere Bedrohung, wie etwa die Sowjetunion, sondern von innen: permanent verdrängte nationale und ethnische Spannungen, soziale Konflikte, die drohende Verelendung breiter Bevölkerungsschichten. All das birgt den Keim für Bürgerkriege und Grenzscharmützel, bereitet verantwortungslosen Populisten den Boden, öffnet der Machtübernahme reaktionärer, ja faschistischer Regime Tür und Tor' (Koydl SZ, 25 May 1991).

15 'Heute haben wir die Chance ... gemeinsam mit den mittel-, ost- und südosteuropäischen Staaten die Einheit Europas zu vollenden. Wir, die Deutschen, haben daran ein ganz besonderes Interesse. Unser Land liegt in der Mitte des Kontinents. Wir haben die meisten Nachbarn und die längsten Grenzen' (Kohl, 12 June 1997, in *Deutscher Bundestag* 13/181: 16224).

16 E.g. Josef Joffe, in SZ, 21 September 1995; Christoph Zöpel, in the German Bundestag 12/50: 4094.

17 'Das Konzept der europäischen Einigung ist und bleibt die einzige wirksame Versicherung gegen Nationalismus, gegen Machtpolitik und gegen Krieg. (Beifall bei der CDU/CSU, der F.D.P. und dem BÜNDNIS 90/DIE GRÜNEN sowie bei Abgeordneten der SPD)' (Kohl, *Deutscher Bundestag* 13/148: 13329).

18 'Die Gründerväter der Europäischen Gemeinschaft hatten nach dem Krieg eine große Vision: unseren Kontinent, der sich jahrhundertelang in Bruderkriegen zerrieben hat, dauerhaft zu befrieden. Die immer engere Integration hat Kriege in Westeuropa undenkbar gemacht und die Grundlage für eine nie gekannten Wohlstand gelegt. ... Das Ende des Ost-West-Konflikts hat uns jetzt die Chance eröffnet, dieses Modell sozusagen auf ganz Europa auszudehnen' (Kinkel, 11 June 1997, *Deutscher Bundestag* 13/180: 16161).

19 Graphical representations of paradigms like these were introduced by Axelrod (1976), comprising 'the policy alternatives, all of the various causes and effects, the goals, and the ultimate utility of the decision maker,' and were the 'concepts a person uses are represented as points, and the causal links between these concepts are represented as arrows between these points' (1976: 5). My graphs are different in at least two important ways. Theoretically, causal beliefs are defined as social phenomena and part as a discursive structure underlying public debate (cf. Shapiro et al. 1988). Methodologically, while Axelrod and his colleagues use positive and negative relationships between concepts, all my arrows represent assertions of positive relationships to avoid the overcomplexity of classical 'cogmap' graphs. To make this strategy feasible, the formulation of concepts has to be adjusted, e.g. the assertion 'democracy makes war unlikely' has to be translated into 'democracy leads to international security.'

20 'Die Osterweiterung, verehrter Kollege Glos, ist nicht nur ein Mittel des Almosens für frühere kommunistische Staaten, sondern sie ist die große Chance, junge Demokratien in Osteuropa zu belohnen und zu festigen (Beifall bei der F.D.P. und der CDU/CSU)' (Haussmann, *Deutscher Bundestag* 13/77: 6737).

References

Adler, Emanuel (1997) Seizing the Middle Ground. Constructivism in World Politics, *European Journal of International Relations* 3 (3): 319–363.

Axelrod, Robert (ed.) (1976) *Structure of Decision* (Princeton, NJ: Princeton University Press).

Baldwin Richard E. and Anthony J. Venables (1995) Regional Economic Integration, in Gene Grossman and Kenneth Rogoff (eds.) *Handbook of International Economics*, III (Amsterdam: Elsevier), pp. 1597–1644.

Batagelj, Vladimir and Matevz Bren (1993) Comparing Resemblance Measures. Extended version of a paper presented at Distancia '92, 22–26 June 1992, Rennes, France.

Baumann, Rainer (2002) The Transformation of German Multilateralism. Changes in the Foreign Policy Discourse since Unification. Paper prepared for presentation at the 43rd Annual Convention of the International Studies Association in New Orleans, 23–27 March 2002.

Baumann, Rainer, Volker Rittberger and Wolfgang Wagner (1999) Macht und Machtpolitik. Neorealistische Außen-politiktheorie und Prognosen über die deutsche Außenpolitik nach der Vereinigung, *Zeitschrift für Internationale Beziehungen* 6 (2): 245–286.

Bulmer, Simon, William Paterson and Charlie Jeffery (2000) *Germany's European Diplomacy: Shaping the Regional Milieu* (Manchester: Manchester University Press).

Burkhart, Ross E. and Michael S. Lewis-Beck (1994) Comparative Democracy: The Economic Development Thesis, *American Political Science Review* 88 (4): 903–910.

Campbell, John L. (1998) Institutional Analysis and the Role of Ideas in Political Economy, *Theory and Society* 27: 377–409.

Checkel, Jeffrey T. (1998) The Constructivist Turn in International Relations Theory, *World Politics* 50 (2): 324–348.

Dahl, Robert A. (1989) *Democracy and its Critics* (New Haven, CT: Yale University Press).

De Witte, Bruno (2002) Anticipating the Institutional Consequences of Expanded Membership of the European Union, *International Political Science Review* 23 (3): 235–248.

Diez, Thomas (1999) *Die EU lesen: Diskursive Knotenpunkte in der britischen Europadebatte* (Opladen: Leske & Budrich).

Doty, Roxanne Lynn (1993) Foreign Policy as Social Construction: A Post-Positivist Analysis of US Counterinsurgency Policy in the Philippines, *International Studies Quarterly* 37 (3): 297–320.

Eberwein, Wolf-Dieter and Matthias Ecker-Ehrhardt (2001) *Deutschland und Polen: eine Werte und Interessengemeinschaft? Die Elitenperspektive* (Opladen: Leske & Budrich).

Ecker-Ehrhardt, Matthias (2002) Alles nur Rhetorik? Der ideelle Vorder- und Hintergrund der deutschen Debatte über die EU-Osterweiterung, *Zeitschrift für Internationale Beziehungen* 9 (2): 209–252.

Ecker-Ehrhardt, Matthias (2004) *Zu Emergenz und Wandel argumentativer Koalitionen- die Integration von Kritik am Beispiel des deutschen Osterweiterungskonsenses. InIIS

discussion paper, vol. 9 (Bremen: Institut für Internationale und Interkulturelle Studien (InIIS), Universität Bremen).
Elster, Jon (1998) Introduction to *Deliberative Democracy*, ed. J. Elster (Cambridge: Cambridge University Press).
Fierke, Karin M. and Antje Wiener (1999) Constructing Institutional Interests: EU and NATO Enlargement, *Journal of European Public Policy* 6 (5): 721–742.
Finnemore, Martha and Kathryn Sikkink (1998) International Norm Dynamics and Political Change, *International Organization* 52 (4): 887–917.
Fleiss, Joseph L. (1981) *Statistical Methods for Rates and Proportions*, 2nd edn. (New York: Wiley).
Friis, Lykke (1998) EU Enlargement and the Luxembourg Summit: a Case Study in Agenda Setting, in A. Wivel (ed.) *Explaining European Integration* (Copenhagen: Copenhagen Political Studies Press).
Gow, James (1999) Security and Democracy: The EU and Central and Eastern Europe, in K. Henderson, *Back to Europe: Central and Eastern Europe and the European Union* (London: University College London).
Gower, Jackie (1999) EU Policy to Central and Eastern Europe, in K. Henderson (ed.) *Back to Europe: Central and Eastern Europe and the European Union* (London: University College London).
Grabbe, Heather (2002) European Union Conditionality and the Acquis Communautaire, *International Political Science Review* 23 (3): 249–268.
Grieco, Joseph M. (1997) Realist International Theory and the Study of World Politics, in M. W. Doyle and G. J. Ikenberry (eds.) *New Thinking in International Relations Theory* (Boulder, CO: Westview Press).
Haas, Peter M. (1992) Introduction: Epistemic Communities and International Policy, *International Organization*, special issue 46, 1: 1–36.
Habermas, Jürgen (1981) *Theorie des kommunikativen Handelns. Band 1: Handlungsrationalität und gesellschaftliche Rationalisierung. Vol, 2: Zur Kritik der funktionalistischen Vernunft.* 2 vols. (Frankfurt am Main: Suhrkamp Taschenbuch).
Habermas, Jürgen (1992) *Faktizität und Geltung: Beiträge zur Diskurstheorie des Rechts und des demokratischen Rechtsstaats* (Frankfurt am Main: Suhrkamp).
Hall, Peter A. (1993) Policy Paradigms, Social Learning, and the State, *Comparative Politics* 25: 275–296.
Harnisch, Sebastian (2001) 'Truth is what works'?, oder 'Was nicht überzeugen kann, das wird sich auch nicht bewahrheiten' – Eine Replik auf Gunther Hellmanns 'Rekonstruktion der Hegemonie des Machtsstaates Deutschland unter modernen Bedingungen'? Fassung vom 18 Oktober 2000: Universität Trier.
Heitzer-Suša, Elke (2001) *Die ökonomische Dimension der EU-Osterweiterung* (Baden-Baden: Nomos).
Hellmann, Gunther (2000) Rekonstruktion der 'Hegemonie des Machtsstaates Deutschland unter modernen Bedingungen'? Zwischenbilanzen nach zehn Jahren neuer deutscher Außenpolitik. Paper read at Vorgelegt auf dem 21. Wissenschaftlichen Kongreß der Deutschen Vereinigung für Politische Wissenschaft in Halle/Saale, 1–5 October 2000.
Hellmann, Gunther (2002) Sag beim Abschied leise Servus. Die Zivilmacht Deutschland beginn, ein neues 'Selbst' zu behaupten, *Politische Vierteljahresschrift*, 43 (3): 498–507.

Hofmann, Jeanette (1993) *Implizite Theorien in der Politik. Interpretationsprobleme regionaler Technologiepolitik* (Opladen: Westdeutscher Verlag).

Hülsse, Rainer (2003) Sprache ist mehr als Argumentation. Zur wirklichkeitskonstituierenden Rolle von Metaphern, in *Zeitschrift für Internationale Beziehungen (ZIB)* 10 (2): 211–246.

Kepplinger, Hans Mathias (2002) Mediatization of Politics: Theory and Data, *Journal of Communication* 52: 972–986.

Kerski, Basil (2001) Die Rolle nichtstaatlicher Akteure in den deutsch-polnischen Beziehungen vor 1990, in W.-D. Eberwein and B. Kerski (eds.) *Die deutschpolnischen Beziehungen 1949–2000: Eine Interessen- und Wertegemeinschaft* (Opladen: Leske & Budrich).

Laclau, Ernesto and Chantal Mouffe (1985) *Hegemony and Socialist Strategy* (London: Verso).

Laffay, Mark and Jutta Weldes (1997) Beyond Belief: Ideas and Symbolic Technologies in the Study of International Relations, *European Journal of International Relations*, 3 (2): 193–237.

Lippert, Barbara (2002) Die EU-Erweiterungspolitik nach 1989 – Konzeptionen und Praxis der Regierung Kohl und Schröder, in H. Schneider, M. Jopp and U. Schmalz (eds.) *Eine neue deutsche Europapolitik? Rahmenbedingungen – Problemfelder – Optionen* (Bonn: Europa Union Verlag).

Lipset, Seymour M. (1959) Some Social Requisites of Democracy: Economic Development and Political Legitimacy, *American Political Science Review*, 53: 69–105.

Luhmann, Niklas (1984) *Soziale Syste,: Grundiß einer allgemeinen Theorie* (Frankfurt a.M: Suhrkamp).

Markovits, Andrei S. and Simon Reich (1997) *The German Predicament. Memory and Power in the New Europe* (London: Cornell University Press).

Maull, Hanns, Sebastian Harnisch and Constantin Grund (eds.) (2003) *Deutschland im Abseits? Rot-grüne Außenpolitik 1998–2003* (Baden-Baden: Nomos).

Moravcsik, Andrew (1996) Federalism and Peace: A Structural Liberal Perspective, *Zeitschrift für Internationale Beziehungen*, 3 (1): 123–132.

Moravcsik, Andrew (2000) The Origins of Human Rights Regimes: Democratic Delegation in Postwar Europe, *International Organization*, 54 (2): 217–252.

Peters, Bernhard (1993) *Die Integration moderner Gesellschaften* (Frankfurt am Main: Suhrkamp).

Przeworski, A. (1998) Deliberation and Ideological Domination, in J. Elster (ed.) *Deliberative Democracy* (Cambridge: Cambridge University Press).

Risse, Thomas (2000) 'Let's Argue': Communicative Action in World Politics, *International Organization* 54 (1): 1–39.

Rittberger, Volker and Wolfgang Wagner (2001) German Foreign Policy after Unification: Theories Meet Reality, in V. Rittberger (ed.) *German Foreign Policy Since Unification* (Manchester: Manchester University Press).

Rucht, Dieter (2002) Rückblicke und Ausblicke auf die globalisierungskritischen Bewegungen, in H. Walk and N. Boehme (eds.) *Globaler Widerstand. Internationale Netzwerke auf der Suche nach Alternativen im globalen Kapitalismus* (Münster: Westfälisches Dampfboot).

Rupp, Michael Alexander (1999) The Pre-Accession Strategy and the Governmental Structures of the Visegrad Countries, in K. Henderson (ed.) *Back* to Europe: Central and Eastern Europe and the European Union (London: University College London).

Sabatier, Paul A. and Hank C. Jenkins-Smith (1999) The Advocacy Coalition Framework: An Assessment, in P. A. Sabatier (ed.) *Theories of the Policy Process* (Boulder, CO: Westview Press).

Schimmelfennig, Frank (1995) *Debatten zwischen Staaten. Eine Argumentationstheorie internationalerSystemkonflikte* (Opladen: Leske und Budrich).

Schimmelfennig, Frank (2001) The Community Trap: Liberal Norms, Rhetorical Action, and the Eastern Enlargement of the European Union, *International Organization* 55 (1): 47–80.

Schimmelfennig, Frank (2003) *The EU, NATO and the Integration of Europe* (Cambridge: Cambridge University Press).

Schimmelfennig, Frank and Ulrich Sedelmeier (2002) Theorizing EU Enlargement: Research Focus, Hypotheses, and the State of Research, *Journal of European Public Policy*, 9 (4): 500–528.

Schneider, Heinrich, Mathias Jopp and Uwe Schmalz (eds.) (2002) *Eine neue deutsche Europapolitik? Rahmenbedingungen – Problemfelder – Optionen* (Bonn: Europa Union Verlag).

Schulz, Winfried (2004) Reconstructing Mediatization as an Analytical Concept, *European Journal of Communication* 19 (March): 87–101.

Sedelmeier, Ulrich (2000) Eastern Enlargement: Risk, Rationality, and Role-Compliance, in M. G. Cowles and M. Smith (eds.) *The State of the European Union. Risks, Reform, Resistance, and Revival* (Oxford: Oxford University Press).

Sedelmeier, Ulrich and Helen Wallace (1996) Policies towards Central and Eastern Europe, in H. Wallace and W. Wallace (eds.) *Policy-Making in the European Union* (Oxford: Oxford University Press).

Shapiro, Michael J., Matthew G. Bonham and Daniel Heradsveit (1988) A Discourse Practices Approach to Collective Decision Making, *International Studies Quarterly* 32 (2): 397–420.

Straubhaar, Thomas and Martin Wolburg (1999) Brain Drain and Brain Gain in Europe: An Evaluation of East European Migration to Germany, *Jahrbücher für Nationalökonomie und Statistik*, 218 (5–6): 574–604.

Tewes, Henning (1998) Between Deepening and Widening: Role Conflict in Germany's Enlargement Policy, *West European Politics* 2 (21): 117–133.

Titschler, Stefan, Ruth Wodak, Michael Meyer and Eva Vetter (1998) *Methoden der Textanalyse* (Opladen: Westdeutscher Verlag).

Torfing, Jacob (1999) *New Theories of Discourse* (Oxford: Blackwell).

Toulmin, Stephen E. (1996 [1975]) *Der Gebrauch von Argumenten*, 2nd edn. (Weinheim: Beltz Athenäum).

van Dijk, Teun A. (1977) *Text and Context: Explorations in the Semantics and Pragmatics of Discourse* (London: Longman).

Waever, Ole (1994) Resisting the Temptation of Post Foreign Policy Analysis, in W. Carlsnaes and S. Smith (eds.) *European Foreign Policy: The EC and Changing Perspectives in Europe* (London: Sage).

Waever, Ole (1998) Explaining Europe by Decoding Discourses, in A. Wivel (ed.) *Explaining European Integration* (Copenhagen: Copenhagen Political Studies Press).

Weise, Christian (2002) How to Finance Eastern Enlargement of the EU. The Need to Reform EU Policies and the Consequences for the Net Contributor

Balance, DIW Discussion Paper 287 (Berlin: Deutsches Institut für Wirtschaftsforschung).

Welzel, Christian and Ronald Inglehart (1999) Analyzing Democratic Change and Stability: A Human Development Theory of Democracy, WZB Discussion Paper FS III 99-202 (Berlin: Wissenschaftszentrum Berlin für Sozialforschung).

Wendt, Alexander (1999) *Social Theory of International Politics* (Cambridge: Cambridge University Press).

Zürn, Michael (2004) Global Governance under Legitimacy Pressure, *Government and Opposition*, 39 (2): 260–287.

13
Media Communication, Citizens and Transnational Risks: The Case of Climate Change and Coastal Protection

Harald Heinrichs and Hans Peter Peters

Introduction

The risk of climate change and its potential consequences for coastal protection are not just issues for professionals such as scientific experts and policymakers. In recent years they have become major public issues: next to the climate protection discourse, the media have started reporting on the risks of climate-related sea level rises, the likelihood of extreme weather events such as storm surges and the inadequacy of existing coastal protection systems. With regard to political decision-making this public discourse is highly relevant: in democracies policymakers need public legitimation as well as the support of their constituencies for collective, binding decisions. Especially in relation to far-reaching, costly decisions like the climate-related adaptation of coastal protection infrastructure where the knowledge is still uncertain, evaluation ambivalent and controversies ongoing, public reaction is critical to success.

With this in mind, we explore in this chapter what we call 'the public risk construct of climate change and coastal protection'. After presenting some general remarks on the challenge of climate change and coastal protection for science, politics and society, we present results from a recent study, which analysed the interaction between experts on climate change and coastal protection and journalists, the media coverage of these issues, and hoe people responded to the coverage. We conclude with an overview of the relationship between media communication, the general public and transnational risks.

The challenge for science, politics and society: climate change and coastal protection

Global climate change has been one of the most prominent transnational risks for the last 15 years. International organizations, governments and NGOs have responded to that challenge and have initiated complex governance processes at the local, national and international levels. The Kyoto Protocol as a first step towards limiting greenhouse gas emissions, for example, has been ratified by 141 nations. Even though the US, the world's biggest polluter, has not joined the international climate protection community, this binding agreement is a significant success for global collective opinion-building and decision-making.

Over the years many different actors, including scientists, policymakers, private business, the media and citizen groups have been engaged in the transnational risk debate. Due to competing knowledge claims, conflicting interests and pluralistic values it is no surprise that there is no unanimity on this highly complex issue. However, as documented by the International Panel of Climate Change Reports with their large-scale peer review, most climate researchers seem to agree that climate change will happen.[1] Nevertheless uncertainty remains about the extent to which human intervention is contributing to climate variability.

The scientific discussion about manmade and natural climate change is especially important because of its potential consequences and its imminent social and economic relevance of mitigation strategies which are highly 'trans-scientific'. Science, values and interests are intertwined, as McCright and Dunlap (2000, 2003) have shown in there exemplary studies on the role of neo-conservative think-tanks in the US. According to their findings since the beginning of the 1990s the conservative 'counter-movement' has systematically – and in the end 'successfully' – instilled a 'climate-sceptical' perspective in the US, based on very few scientists, which aims at the 'de-problematization' of human-induced climate change.

Despite these science-based policy controversies on the reality and causes of climate change, a parallel discourse has developed which has climate change as its starting point and focuses on the potential consequences. Extreme weather, such as storms, floods and heat-waves, melting glaciers and rising sea levels are the predicted consequences (Turner et al. 1995; Schirmer and Schuchardt 2005). Based on computer simulations, scenario techniques and analyses of current weather-related catastrophes, this research aims at producing anticipatory knowledge to identify adaptation strategies proactively.

Coastal zones are of special interest. Because of potential sea level rises, storm tides and floods may become stronger and more frequent. Since almost half the world's population live on or near the coast, and therefore highly valuable infrastructure is concentrated in these regions, preventative adaptation strategies in coastal protection and coastal zone management are important to limit vulnerability to climate change (Sterr et al. 1999, 2000). Studies have analysed the risk to coastal zones. And specific research and development programmes, such as the EU programme on integrated costal zone management (IKZM), aim at sustainable strategies, which integrate different challenges to coastal zones, including potential climate change consequences, in anticipatory decision-making.[2] Even though early efforts have been made to analyse climate change consequences more systematically and to identify possibilties for adaptation, climate change and mitigation continue to dominate the scientific and political discourse (Weingart et al. 2002; BMBF 2003).

In sum we can say that there are significant discourses in science and politics about the transnational risk of global climate change and its potential consequences. Many activities at different levels of research and decision-making have commenced during the last two deades. The issue is widely considered to be one of the most urgent problems humanity is currently facing. One British climate researcher, Sir John Houghton, has even compared the risk of clmate change with weapons of mass destruction: 'Like terrorism, this weapon knows no boundaries. It can strike anywhere, in any form – a heat wave in one place, a drought or a flood or a storm surge in another.'[3] Basic socio-economic structures of modern industrial societies are seen as part of the problem, and large parts of the world may be at risk. Because of the scale of the problem it is not surprising that the topic is considered not just as a matter of technical management, which could be solved by experts and policymakers alone. The professional discourses in science and policymaking were accompanied from an early stage by public communication. Climate change has thus become a public issue, with relevance for potentially every citizen (Weingart et al. 2002).

Accordingly social sciences have not only analysed the climate change sciences, scientists and knowledge production, as well as the political decision-making processes (Boehmer-Christiansen 1994; Stehr and von Storch 1995; Bray and von Storch 1999; Clark and Dickson 1999; Beck 2004; Fogel 2004), but the (risk) perception of citizens too. Alongside a better understanding of the scientific construction of 'climate change' and the development of climate policies, the question of how lay people interpret global climate change has been an important research question since the late 1980s.

Numerous studies have analysed people's understanding, risk perception and opinions regarding problem solutions with regard to climate change – both its causes and consequences (e.g. Bell 1989; Wiedemann 1992; Kempton 1997; Dunlap 1998). The studies deal with representative surveys – national and international comparative – up to psychological analysis of mental models to compare expert and lay people's representations of climate risks. The rationale of these studies is to understand better how the people perceive and construct this issue, because public opinion – as support for and/or pressure on political decision-making – is seen as essential to the implementation of effective (and not only symbolic) climate policies. The most important results of the national and international comparative studies are these:

- Many in industrialized and developing countries have heard about climate change, but the level of concern varies between countries.
- The cognitive construction of the issue is often mixed with other global environmental problems, such as the ozone layer and air pollution.
- In industrialized countries there is a tendency for a time–space distant perspective, climate change in the future and in developing countries evokes more concern than the local, near-future situation. In developing countries it's the other way round.

Even though climate change is a transnational and 'glocal' risk because globally distributed causes (emission of greenhouse gases) influence the global climate, which in turn leads to local/regional consequences – for example, for coastal protection – most of the 'first-generation' studies on climate change perception focused on the general level, without reference to a specific local situation.

With the dawn of climate impact research the link between global and local/regional levels came into focus. At the German North Sea coast, for example, Volker Linneweber et al. (2001), in their study of the potential climate change consequences for the island of Sylt, analysed the semantic associations between local coastal protection and global climate change. Their results show that coastal protection, compared with other problems on the island, is of low priority despite extensive media coverage. Likewise climate change does not rank high on people's problem agenda. Moreover the study shows that people prefer options for climate mitigation over climate adaptation, such as the extension of coastal protection infrastructure, which may interfere with the demands of tourism. Place-based studies which link global climate change with potential local consequences have added importantly to the discourse on risk perception of climate change.

Since abstract risks, such as climate change and its potential consequences, are (at first) accessible only by scientific methods and interpretations, especially by anticipative computer simulation and scenario techniques, people's representations of the risk are not developed via direct perception. In science-based modern societies risk perception and the emergence of risk awareness are highly influenced by the media (e.g. Peters 1994; Heinrichs 2003). Mass media play a major role in creating a public sphere that serves as an arena for the process of knowledge distribution and opinion-building. Consequently, media communication on climate change has been an important research topic since the end of the 1980s.

While there are several other arenas involved in the management of collective problems, the public arena has a number of distinctive features:[4]

1. Communication processes taking place in the public arena help new concerns, actors and ideas to enter the political agenda, urging social and political institutions to react ('agenda-setting'). They hence stimulate socio-political innovation in order to adapt to a changing environment.
2. Public communication helps to create an issue-specific cultural context ('issue culture'), providing e.g. shared knowledge and problem frames serving as a common reference basis for all involved actors ('issue framing').
3. Public communication links political processes on the political 'stage' – involving mainly social actors and political decision-makers – to the broader population. This is a necessary (although not sufficient) precondition of political participation.

The public sphere is not established by mass media alone but by a communication system characterized above all by the interaction of journalism, social actors and media audiences. Scientific experts and institutions, politicians and political institutions, NGOs and citizen action groups implement different kinds of public relations strategies and feed the media with information, interpretations, demands, suggestions, criticism, metaphors and catchphrases. Furthermore, the media public has to be taken into account. First, the anticipated interests of the media audience are an important factor for decisions made by social actors as well as by journalists. Second, reception of media coverage is a highly selective and interpretative process which is not dominated by content but influenced, for example, by experiences, heuristics, values, interests, preconceived attitudes and socio-economic characteristics of the media audience.

Based on this general knowledge and understanding of the media as a social system with specific production and reception processes, many studies have been conducted on climate change and the mass media (e.g. Bell 1989, 1994; Mazur and Lee 1993; Wilkins 1993; Clark and Dickson 1995; Peters and Sippel 1998). Some studies have analysed the media content in order to gain an overview of the public discourse with regard to relevant actors, events, topics, claims and interpretations. Others have observed from a critical perspective the media as 'actor', which influences the availability of public information and interpretation on climate change and in this way the opinion-formation of the people. These studies have criticized, from a specific climate research point of view, an unfounded catastrophe thesis, one-sided selection by experts as well as in part inaccurate coverage or inadequate contextualization. Both types of study, however, confirm the special role of scientific experts as journalistic sources, even though in the meantime sources of the political-administrative system have gained in relevance (Trumbo 1996). The high credibility of experts, who mostly appear as warning signals in the media, is seen as a specificity of this issue (Wilkins 1993). In relation to this research perspective, Allan Bell has analysed the differences between media coverage and public opinion. His studies show that people are more concerned about climate change than the media coverage would suggest. Moreover, he found that people conflated different climate-related problems and mentioned with regard to the ozone layer, selective causes, which are connected to their daily life. Meanwhile the media present a broad range of possible causes.

Taking into account these heterogeneous studies, we analysed in a large-scale empirical project (2001–4) the communication and reception processes among experts, journalists and media consumers in order to gain an understanding of the public risk construction of climate change and coastal protection on the German North Sea coast.[5] In the next section we focus on the results of media representation and general awareness.

Case study: media representation and general awareness of climate change and coastal protection at the German North Sea coast

A differentiated analysis of the public and individual risk representation raises a broad range of questions: How are climate change and coastal protection represented in the media and what do the people think about the issue? What kind of interpretations, knowledge claims and evaluations are presented, and how do the people interpret them? Which risks are discussed in the coverage and how do the people assess them? What options for action are proposed in the media and what do people think about coping strategies? Who are the main actors in the media discourse and what do people think about different actors?

In our study we combined two methods – media analysis and surveys – in order to analyse media representation and general awareness of climate change and coastal protection with a coherent research design.

In order to gain a detailed understanding of the mass media-based symbolic environment we analysed newspapers, magazines, radio and television programmes. The aim was to set up a number of articles similar to the actual media environment of the local population at three locations on the North Sea coast: Bremen, a large city; Wilhelmshaven, a small city; and Wangerland, a rural area. For these locations we have monitored the relevant local, regional and the most important national media. The identification of articles was guided by the following thematic criteria:

Climate change:

- Description/causes of climate change
- Mitigation of climate change
- Consequences of climate change
- Adaptation to climate change
- Climate research
- Political debate about climate change

Coastal protection:

- Coastal risks because of sea level rise and extreme weather events, e.g. storm tides
- Coastal protection: status, repair, planning of dikes and technical infrastructure
- Coastal protection research
- Political debate about coastal protection

Between 1 September 2001 and 28 February 2003, we selected 1,176 articles. Even though we invested a lot effort in monitoring radio and television programmes, the analysis shows that 85 per cent of the media products are newspaper and magazine articles (Table 13.1).

The quantitative content analysis of the articles was guided by a code book and conducted by students, who were trained in several sessions, which guaranteed sufficient reliability. Based on our research perspective the thematic analysis focused on: information, statements, evaluation regarding the general topic, actors, risk existence, risk causes, risk acceptance, risk responsibility and risk coping. This enabled a differentiated analysis of the media coverage.

Awareness of climate change and coastal protection was analysed in a preliminary study. In survey sessions in the three locations (Bremen, Wilhelmshaven and Wangerland) we combined a standardized questionnaire with a quasi-experimental input to analyse the cognitive processes in the reception of media articles. This approach had been successfully employed in earlier studies (Peters, 1999).[6] In each location approximately 60 people, selected at random, were interviewed with standardized questions about pre-knowledge, opinions, attitudes, risk perception and evaluation as well as personal characteristics to measure variables such as environmental awareness, trust in institutions, attitudes towards coastal protection, perceptions of climate and coastal

Table 13.1 Composition of media sample for content analysis

			Subject			
			Climate change	Coastal protection	Both	Total
National media	Print	Article	415	5	12	432
	Radio	News	46	1	0	47
		Magazine	54	2	2	58
	TV	News	21	4	0	25
		Magazine	34	2	0	36
		Documentary	6	1	1	8
Regional/local media	Print	Article	285	205	18	508
	Radio	News	5	0	0	5
		Magazine	5	10	12	27
	TV	News	0	0	0	0
		Magazine	9	8	10	27
		Documentary	2	0	1	3
Total			882	238	56	1,176

risks, anxiety, evaluation of media coverage and willingness to act, as well as socio-demographic characteristics. Due to the relatively small number (180 survey participants) and the low response rate in our random sampling approach, representativeness is limited. But the results of the survey questions regarding climate change, which were included in a representative national survey too, show fairly good agreement. Our interviews assess the perceived risks of climate change to be only slightly higher than those of the respondents in the national study.

Based on the data of the media study and the standardized questionnaire of the preliminary study, we will show the differences and similarities between media representation and people's awareness of the risks of climate change and coastal protection, options for mitigation and adaptation and trust in institutions and responsibilities.

Risks of climate change and coastal protection

We analysed the representation of 'climate risks' and 'coastal risks'. The two are potentially connected, because climate change is expected to cause sea levels to rise and the likelihood of extreme weather events, such as storm tides and flooding. In the media coverage, however, they are rarely linked: just 6 per cent of the 1,176 media products integrate the issues. But the coverage on climate change and coastal protection differs significantly between local/regional and national media. National media report very rarely (4 per cent) the risks to coastal zones in general and climate change-related coastal risks. In the local/regional media 46 per cent of the selected contributions are about coastal issues. Coastal protection is mainly a regional issue. Interestingly, almost 10 per cent of the regional articles/broadcasts do link climate change and coastal protection discourses (Table 13.2).

Table 13.2 Topics of media stories

	All media	National media	Regional media
Only climate change	74.1%	94.2%	52.8%
Only coastal protection	18.6%	2.0%	36.3%
Climate change and coastal protection	5.5%	1.8%	9.5%
Neither climate change nor coastal protection	1.7%	2.0%	1.4%
	100.0% (n = 1.176)	100.0% (n = 606)	100.0% (n = 570)

Table 13.3 Alarming vs. reassuring tone of media coverage

	Climate change		Costal risks	
	Title	Text	Title	Text
Alarming	35.9%	48.9%	34.2%	37.0%
Reassuring	6.2%	5.9%	15.5%	16.9%
Without/no clear tendency	58.0%	45.3%	50.3%	46.1%
	100.0%	100.0%	100.0%	100.0%
	(n = 937 stories)		(n = 284 stories)	

Risks are expectations about possible harm. Therefore we analysed what possible harm is represented in the media coverage and to what extent this is interpreted as dangerous, not very dangerous or not dangerous at all. At first we investigated whether the article, including the headline, in general was alarmist or reassuring. Second, we identified and classified risk-related statements in the coverage. As Table 13.3 shows, warnings regarding the risks of climate change predominate. In the (regional) coverage there are slightly more articles/broadcasts reassuring that the coastal protection system is functioning well.

The media communicate a broad spectrum on possible climate- and coast-related damage. This is defined on different levels: e.g. climate change itself is represented as 'damage' as well as potential consequences such as melting glaciers or sea level rises; in this context secondary consequences – casualties, health risks, economic losses – are discussed. Since climate change is predominantly reported in national media and coastal protection in local/regional media, concrete coastal risks like storm tides and floods are significantly more represented in local/regional media.

With regard to all reported potential damage the media mostly publish statements which confirm the risks and present it as dangerous. Less than 3 per cent of all statements contest or relativize the existence of climate change risks. In Germany mass media climate change is communicated as a serious risk; sceptical voices are rarely raised (Table 13.4).

Beyond the identification of risk statements in the media coverage we analysed how cause-and-effect relationships are presented. By far the most frequently mentioned cause-and-effect relationship is manmade CO_2 emissions as the principal cause of the greenhouse effect. Natural causes, such as geological processes, are discussed far less often. The greenhouse effect itself is seen as the cause of extreme weather, damage

Table 13.4 Risks mentioned in the media coverage

	All media	National media	Regional media
Climate change/greenhouse effect	19.3%	25.8%	12.5%
Melting of glaciers	7.4%	10.0%	4.7%
Sea-level rise	6.1%	6.1%	6.2%
Extreme weather events	12.9%	12.7%	13.1%
Floods, storm tides (sea)	8.6%	2.7%	14.7%
Flooding (inland)	10.2%	9.3%	11.1%
Natural hazards	1.7%	2.3%	1.1%
Flooding of islands (Tuvalu)	1.3%	1.6%	0.9%
Shift of vegetation zones	5.9%	7.5%	4.2%
Material damage, infrastructure, agriculture	10.9%	7.0%	14.9%
Casualties, health damage, suffering	7.1%	5.0%	9.3%
Damage to ecosystems, biodiversity	7.3%	8.9%	5.6%
Other/no specific damage	1.4%	1.2%	1.6%
	100.0% (n = 1.111)	100.0% (n = 561)	100.0% (n = 550)

Table 13.5 Cause-and-effect relationships of climate change

	Number of statements	Cause	No cause
Anthropogenic CO_2 → green house effect	116	81.9%	5.2%
Traffic/industry → green house effect	22	95.5%	0.0%
Geological processes → green house effect	15	80.0%	0.0%
Greenhouse effect → extreme weather event	61	85.2%	3.3%
Greenhouse effect → ecosystems	47	89.4%	0.0%
Greenhouse effect → melting glaciers	44	72.7%	4.5%
Greenhouse effect → vegetation zones	21	76.2%	0.0%
Melting glaciers → sea-level rise	24	83.3%	4.2%

to the eco-system, the melting of glaciers and shifting vegetation zones. The melting of glaciers is reported as the cause of rising sea levels. The identified cause-and-effect relationships indicate that in the German media discourse climate change is presented predominantly as a man-made risk which has catastrophic consequences for the environment and society (Table 13.5).

We have explored the risk awareness of the coastal population who live in this media-based symbolic environment using several questions in our survey. Approximately 87 per cent of the interviewees are 'very convinced' or 'fairly convinced' that the climate will change, 65 per cent are 'not convinced' or 'not convinced at all' that Germany can cope with the consequences, but 57 per cent still think that climate change can be prevented. These results are in line with other studies undertaken in Germany, which identified a high risk awareness regarding climate change among the German population (Table 13.6).

In order to characterize risk representation more precisely we used four opposing adjectives in the questionnaire, to find out the semantic associations regarding climate change, the general risk of storm tides and the personal risk of storm tides. For both risks of climate change and the general risk of storm tides, the interviewees strongly associated the characteristics 'dangerous', 'likely', 'extensive damage' and 'uncontrollable'. Compared with that, the personal risk of storm tides is assessed as

Table 13.6 Conviction of the population that climate change is real and that it can be mitigated

Most climate researchers predict a warming of the global atmosphere. They expect, for instance, a rise in sea levels as well as a shift in climate zones. How strongly are you convinced that the predicted climate change will happen?

	Interviewees	National survey*
Completely convinced	37.4%	27%
Fairly convinced	49.5%	50%
Not very convinced	13.2%	20%
Not convinced at all	0.0%	3%
	100.0%	100%
	(n = 182)	(n = 2.361)

* Representative national survey 2002 (Grunenberg and Kuckartz 2003).

Most climate researchers assume that climate change is caused by human activities, mainly by burning coal and Oil. How strongly are you convinced that climate change can be stopped by adequate measures?

	Interviewees	National survey*
Completely convinced	12.0%	9%
Fairly convinced	44.8%	41%
Not very convinced	38.3%	44%
Not at all convinced	4.9%	6%
	100.0%	100%
	(n = 183)	(n = 2.361)

* Representative national survey 2002 (Grunenberg and Kuckartz 2003).

Table 13.7 Risks of climate change – dangerous, likely, great damage, uncontrollable

	Climate change	Storm tides in general	Storm tides personally
Dreadful vs. not dreadful	2.9	2.8	4.0
Not likely vs. likely	5.0	4.8	3.7
Great damage vs. little damage	3.1	3.0	3.7
Controllable vs. uncontrollable	4.7	4.4	4.1
Risk indices	3.7	3.4	0.0

Basis: n = 183
The associations of the opposing adjectives were rated with a 7-step scale

significantly less dangerous (Table 13.7). The phenomenon that risks are assessed lower for oneself than for the general population is called in risk research 'unrealistic optimism' (Weinstein, 1989). In our case study, however, not all people on the North Sea coast bear the same risk in the event of a strong storm tide.

Next to the analysis of risk perception we explored statistical correlations and associations between the risk-related indices and sociodemographic variables, personal characteristics and political-cultural variables. Due to the relatively small size of our sample (n = 180) only strong correlations are significant. Hence we have found few significant correlations. Consistent are correlations between risk perception and age as well as risk perception and environmental concern: older interviewees assess the risk as lower than younger ones, and those with high environmental awareness assess the risk as greater than those with lower environmental concerns. Other studies show that people put the issue of climate change within the broader context of general environmental problems.

We found significant gender-specific differences in risk perception only for the belief that global climate change is imminent: men are more convinced than women. This is surprising because, as other studies have confirmed, women tend to evaluate risks as higher than men and have higher scores in environmental concern. The latter we found in our study too.

As the media analysis and the survey show, global climate change and its potential consequences for the German North Sea coast are interpreted by media and citizens as a serious risk. In German media coverage there is a very high consonance about manmade climate change. And within the (coastal) population there is a high risk awareness about climate change and coastal protection. But more than half the interviewees remain optimistic that climate change can be prevented, and the personal risk of storm tides is seen as not particularly threatening. Keeping in mind that the media discourse on mitigation is much stronger than the coverage on adaptation in general and coastal protection specifically, it may not be surprising that risk awareness is more focused on the general risk of climate change and the issue of mitigation than on personal threats and adaptation.

Options for mitigation and adaptation

After discussion of risks of climate change, the issue of coping is the second most reported topic in the media coverage. There are articles/broadcasts about mitigation strategies to prevent or reduce the greenhouse effect, and adaptation strategies to prepare society for the consequences of climate change. The quantitative distribution shows that mitigation is dominating the coping discourse: 62 per cent of the measures are related to reducing the greenhouse effect, and only 23 per cent focus on adaptation to climate change impacts. Moreover, the representation of climate-related coastal protection is almost entirely a regional/local issue (Table 13.8).

The five most reported coping measures are:

1. *International agreements for climate protection* – these are dealt with in relation to mitigation by the reduction of manmade greenhouse gas emissions. In 45 per cent of the statements, international agreements are commented on positively, in 20 per cent critically. International agreements are the most often discussed mitigation measure. But compared to the other measures, the approval is lowest. The most important aspects were 'necessity' and 'efficacy', followed by 'political enforceability', 'controversy' and 'socio-economic consequences'.
2. *Technical innovation* – e.g. improved energy efficiency or renewable energy – is supported strongly in the media coverage and hardly ever criticized. Next to necessity and efficacy costs and ecological and socio-economic consequences are most often discussed.
3. *Improvement of coastal protection infrastructure (dykes etc.)* – this is almost exclusively reported in regional/local media. These measures have high approval (67 per cent) and receive virtually no criticism (4 per cent). The most important aspects are again necessity and efficacy, followed by cost. The discussion on dykes, etc. is the only one of the five most often discussed coping measures which focus specifically on adaptation and not on mitigation.
4. *National political measures* – reported in the media coverage are related almost entirely to the issue of mitigation. As with international

Table 13.8 Options for mitigation and adaption

	All media	National media	Regional media
Climate protection			
National political measures	14.1%	18.6%	8.6%
International agreements for climate protection	28.2%	34.7%	20.5%
Social innovation (e.g. Agenda 21)	3.6%	4.5%	2.5%
Techonical innovation	16.2%	21.1%	10.4%
Adaption			
Change in agriculture	0.4%	0.8%	0.0%
Change in technical infrastructure	0.8%	0.8%	0.7%
Coastal protection – offensive			
Dyke reinforcement/maintenance	14.2%	1.7%	29.1%
Massive improvement of coastal protection	0.2%	0.0%	0.5%
Coastal protection – defensive			
Retreat/land use change	1.0%	0.2%	2.0%
Ecological restoration/retention areas	0.9%	0.2%	1.7%
Organization of coastal protection			
International agreement on coastal protection	0.2%	0.2%	0.2%
Organization dyke construction and maintenance	0.6%	0.0%	1.2%
Flood protection (inland) – offensive			
Dykes/flood protection infrastructure	0.9%	0.0%	2.0%
River regulation by technical measures	0.6%	0.2%	1.0%
Flood protection (inland) – defensive			
Retreat/land use change	0.8%	0.6%	1.0%
Naturalization/retention areas	1.7%	1.7%	1.7%
Organization of flood protection (inland)			
National/international aggreements on flood protection	0.1%	0.0%	0.2%
Organization of flood protection	0.1%	0.2%	0.0%
Disaster management			
Disaster warning	0.8%	0.4%	1.2%
Disaster management	2.5%	0.8%	4.4%
Regulation of damage	2.6%	1.7%	3.7%
Other			
Other measures	7.8%	9.5%	5.7%
No specific measure	1.8%	2.1%	1.5%
	100.0%	100.0%	100.0%
	(n = 889)	(n = 484)	(n = 405)

agreements approval is somewhat lower (46 per cent) and disapproval somewhat higher (14 per cent) than other measures. Again, the most important aspects in the comments are necessity and efficacy, followed by costs, political enforceability, socio-economic consequences and controversy.

5. *Social innovation (e.g. Agenda 21)* as a coping measure for climate change, its impact is represented only occasionally in the media coverage. But, if mentioned, is commented on very positively (78 per cent);

criticism is rarely expressed (9 per cent). Again, necessity and efficacy are the aspects most often mentioned.

In sum, media coverage on coping measures – especially at the national level – is mainly about mitigation. A considerable discourse on adaptation takes place only in regional/local media.

In order to assess the awareness of the interviewees regarding coping measures and their degree of support, we asked them to react to six statements with approval or disapproval on a five-step rating scale. One should not overvalue willingness to act; nevertheless they indicate the extent to which measures are evaluated and find general support. In all six items the distribution of the answers shows general acceptance and support for measures regarding mitigation and adaptation. The highest degree of approval is expressed for the statement that coastal protection systems should be improved with regard to climate change (90 per cent). Compared to this measure the approval for an eco tax was much lower (50 per cent) (Table 13.9).

Table 13.9 Survey results: mitigation and adaption

	Completely agree	Partly agree	Neither agree nor disagree	Partly disagree	Completely disagree	
Because of climate change coastal protection must be improved, even though it will be expensive	44.8%	44.3%	8.7%	2.2%	0.0%	100.0%
The predicted climate change is no reason for me to make voluntarily changes to my lifestyle	1.6%	10.9%	25.1%	45.9%	16.4%	100.0%
Germany should not take the lead in climate policy, but wait for the international solutions	8.7%	8.7%	11.5%	39.3%	31.7%	100.0%
I am willing to pay higher prices for products, which are less harmful to the climate	27.3%	52.5%	16.4%	3.3%	0.5%	100.0%
The climate change does not currently justify coastly improvements of dykes and other coastal infrastructure	3.3%	17.0%	14.8%	45.6%	19.2%	100.0%
I think it is good that the government uses the eco-tax to encourage the population to adopt energy-saving behaviour, in order to protect the climate	24.7%	26.9%	23.1%	14.8%	10.4%	100.0%

Basis: n = 183.

In order to get a differentiated understanding of awareness regarding mitigation and adaptation we searched for statistically significant correlations between expressed support for measures and potential predictor variables. We found the following patterns. Willingness to support measures increases with:

- higher degree of environmental concern;
- higher assessment of the magnitude of the risk;
- higher assessment of political efficacy, i.e. the expected efficacy of individual and political action;
- greater tolerance of ambiguity, which means the personal competence to handle uncertain situations.
- higher education which is higher for women than for men.

The correlation between willingness to support measures and environmental concern, risk perception, education or gender are in line with other studies. The aspects 'political efficacy' and 'tolerance of ambiguity' add two interesting points: especially in the case of climate change and its potential impacts, where cognitive uncertainty and normative ambiguity are central characteristics, the question of acting under uncertainty and expectations on the efficacy of measures gain in relevance.

The results of the media analysis and of the survey show that there is basic awareness and support for measures as well for mitigation and adaptation. The mitigation discourse, which dominates the national discourse, is sceptically commented on in the media and by the people, especially with regard to important actors such as 'international politicians'.

Adaptation and more specific coastal protection is above all a regional/local issue and is positively assessed by the media and the interviewees.

Trust and responsibilities

In social scientific risk research trust in risk producers and managers is seen as an important factor for risk perception and risk acceptance. In the case of climate change there are no single risk producers – industrial society as a whole is held responsible. And the risk of storm tides for coastal protection is seen as a 'natural risk' and not one that is socially 'produced'. For climate change and coastal protection that means: public and individual trust in the institutions that govern the risk – science, business, politics, administration – is a key factor in the understanding of the societal response to this risk. One can expect that high trust in risk

Table 13.10 Sources of statements about risk in the media coverage

	All media	National media	Regional media
Politics/administration	12.8%	11.9%	13.6%
Business	0.8%	0.9%	0.7%
NGOs: environmental groups, etc.	2.7%	2.7%	2.7%
Science and technology	44.9%	51.9%	37.8%
Other sources	3.3%	1.8%	4.9%
Multiple sources	2.9%	2.7%	3.1%
No Sources	32.6%	28.2%	37.1%
	100.0%	100.0%	100.0%
	(n = 1.111)	(n = 561)	(n = 550)

management should be related to lower risk concern and higher acceptance of proposed measures and coping options (Table 13.10). In the following we describe how actors and institutions are presented in the media with regard to trust and responsibility, and then what our interviewees think about the trustworthiness of responsible institutions.

In the media coverage about two-thirds of all statements can be assigned to primary sources; the statements are based on actors' opinions and referred to or quoted by journalists. The quantitative distribution of primary sources differs between the risk and coping discourses. In the media coverage on risks of climate change 'science and technology' are the dominating source for the journalists, and 'politics and administration' the most important sources in the coverage on coping measures (Table 13.11).

In order to find out to what extent actors/institutions are presented in the media as competent and trustworthy, we classified how the actors and their statements/opinions are represented in the coverage. We constructed an index ranging from +1 (positive evaluation) to −1 (negative evaluation) with 0 as an ambivalent or neutral evaluation. The results show that the media do not present the primary sources and actors negatively. On the contrary, most actors have positive index values, especially 'science & technology'. Only 'politics & administration' is presented more negatively. But a closer look shows that negative evaluation is caused by the actor groups 'foreign politics, international politics and politics in general', mainly due to the critical evaluation of the US's climate policy. Regional German politics and especially the administration are represented more positively.

We explored the interviewees' trust in institutions on two levels: generally, and specifically for coastal protection. The general question for trust

Table 13.11 Sources and statements about coping measures in the media coverage

	All media	National media	Regional media
Politics/administration	34.0%	27.3%	42.0%
Business	4.5%	6.4%	2.2%
NGOs: environmental groups, etc.	7.4%	8.1%	6.7%
Science and technology	19.3%	21.5%	16.8%
Other sources	3.1%	2.3%	4.2%
Multiple sources	13.2%	17.6%	7.9%
No sources	18.4%	16.9%	20.2%
	100.0% (n = 889)	100.0% (n = 484)	100.0% (n = 405)

in institutions shows that 'science' is by far the most trusted institution (93 per cent), followed by the 'media', 'administration', 'legislative & executive', 'industry' and finally 'political parties'. This general trustworthiness can be interpreted as the 'public image' of these institutions. Depending on the particular issues and problems, the trust may vary. The extremely positive image of science, for example, does not mean that in particular cases science would not be criticized. Nevertheless it gives an idea of the general reputation of actors and institutions, which may influence their standing in governing the risks of climate change and coastal protection.

We analysed trust in the institutions responsible for coastal protection, for current and future coastal protection (under conditions of climate change). On a seven-step scale the interviewees were asked to express their trust (low values = high trust; high values = low trust). As the median indicates the interviewees trust the coastal protection system at present and – a little less – under future climate change conditions. It is interesting to note that trust is greatest in the rural area; this could be explained by the greater familiarity with coastal protection infrastructure and responsible actors there.

Bearing in mind the (prior) hypotheses that high trust in risk-regulating institutions is related to lower risk concern and higher acceptance and support of coping measures, we searched for correlations between the trust and risk indices. The results confirm that greater trust in coastal protection systems correlates with lower risk concern. The direction of the relationship, however, remains unclear. Possibly, risk awareness and trust in climate change and coastal protection are interdependent elements of the same semantic syndrome: e.g. people may have high trust in coastal protection because they perceive the risks to be not catastrophic, and do not perceive risks as dreadful because they trust risk management. Compared to this the correlation between trust and coping measures is counter-intuitive: higher trust in coastal protection is correlated with lower willingness to support action. Trust in coastal protection seems to decrease the belief that further risk management activities are necessary. The relationship between general trust in institutions and risk indices shows a positive correlation between trust and the ability to cope with climate change: interviewees with higher trust in institutions are more optimistic that climate change can be handled. This result indicates that people who generally trust societal institutions have a stronger belief that risks can be solved within the current institutional system.

The results of the media study and the survey sketch a differentiated picture about the public and individual attribution of trust to actors and institutions. Science and technology, the most important journalistic source for risk discourse, is the most trusted institution – in the media coverage and in the general opinion of the interviewees. The most important journalistic source for the coping discourse is 'administration & politics'. Administration is still fairly well evaluated; whereas politics, especially foreign/international politics and party politics, are trusted less. Finally, we can state that trust and risk concerns are closely related.

Outlook: media communication, citizens and transnational risks

In this chapter we have analysed the public risk construct of climate change and coastal protection. The public and individual representation of the issue is essential to the policymaking process: in democracies public policy has to be legitimated by the people. Thus, climate change as one of the biggest transnational challenges of the current century is not only an issue for professional risk managers. After more than two decades of media discourse on climate change the study shows that climate change and its expected consequences are embedded in public opinion in Germany.

In the national media climate change is reported as essentially manmade and over half the interviewees believe that climate change is real. Accordingly, the mitigation discourse is dominant. But even though over half of the people are still optimistic that climate change can be prevented, their trust in international/foreign politics as responsible institutions is low. This is in line with the media coverage: they comment on statements from international/foreign policy compared to other sources, especially science, fairly negatively. Coastal protection, on the other hand, is more discussed on the regional/local level. Here more actors/institutions from 'administration & policy-making' are represented in the media. Administration is trusted and the interviewees trust the current coastal protection system and – only a little less – the future coastal risk management under climate change conditions too. But only a minority of the media coverage at the national and local levels is on adaptation issues and more specifically on the relation between climate change and coastal protection. The study shows moreover that climate change is related to the broader environmental awareness, and that risk concerns are connected with trust in institutions.

At the beginning of this chapter we argued that the media play an important role for the communication of abstract risk issues such as climate change. The results presented here indicate that public and individual representation tend to be in the same direction. Since climate change has been in the public domain for some time now, we can observe a co-formation of media representation and citizen awareness. Despite complex communication and reception processes, with construction and reconstruction of meaning, on individual levels, which we have analysed in our project but not presented here, we can state that the media and general public have developed similar representations of climate change and its potential impacts. Bearing in mind the IPCC reports, which characterize climate change as a serious global risk, we can state that in the case of climate change the science-based perspective has reached the public realm in Germany.

However, the dominating representation of mitigation in the media and the optimism of more than half the interviewees that climate change can be prevented on the one hand, and the critical comments of the media regarding international politics and the mistrust in politics of the interviewees on the other, may reflect a societal problem in handling climate change, which cannot be solved easily. Moreover, if the assessment of climate and climate change research, as summarized by the IPCC, is right, there is a growing need for more concrete public communication on adaptation alongside mitigation. In the discourse on adaptation other actors and levels of action gain in relevance.

For the societal resonance to abstract, global risk issues like climate change media communication is essential. More studies, especially in intercultural comparative perspective, are needed, in order to understand better how in public communication processes culture-specific risk constructs arise. Our study reveals exemplarily how this complex, expertise-based issue has become culturally embedded in public communication processes in Germany. In the US, for example, the public risk construct of climate change is quite different: expertise on the topic is differently evaluated in public discourses and large parts of the population seem to be less worried about climate change. But 'successful' public communication is a prerequisite only for 'successful' societal risk management. Concrete problem solutions have to be developed and adopted in other social arenas: politics, business, civil society, private households. Next to traditional instruments of hierarchical political steering (laws, etc.), economic market incentives, and informative-educational methods, participatory approaches seem to offer new creative ways to stimulate social learning and transformation processes (Heinrichs 2005). The opportunities and limits of this procedure are again commented on and disseminated by media communication.

Notes

1 See for IPCC reports: www.ipcc.org.
2 Literature and information on ICZM in Europe: http://europa.eu.int/comm/environment/iczm/.
3 *Guardian*, 28 July 2003.
4 Regarding the role of the media for science, technology and environment, see e.g. Peters (1994); Brand et al. (1997); Luhmann (1986, 1996); Weingart (2001).
5 This was part of the joint project, 'Climate change and preventative risk and coastal management on the German North Sea coast'. It was conducted by an interdisciplinary team of researchers and sponsored by the German Ministry of Research and Education (Schirmer and Schuchardt 2005).
6 Having answered the preliminary questionnaire, the participants were asked to read four articles about climate change and/or coastal protection and to verbalize after each article their thoughts. The verbalized cognitive reactions were recorded. This so-called 'thought-listing-technique' allows an idea of the individual reception processes. After this quasi-experimental input we measured with standardized questions the extent to which risk perception and evaluation have been modified by the articles compared to the answers before reading them. In this chapter we focus on the results of the standardized questionnaire. The cognitive reaction within the reception processes and the analysis of the impact of media coverage on opinion building will be published in a different paper.

References

Bell, Allan (1989) *Hot News. Media Reporting and Public Understanding of the Climate Change Issue in New Zealand. A Study in the (Mis)Communication of Science* (Wellington, Australia: Department of Linguistics, Victoria University).

Bell, Allan (1994) Media (Mis)communication on the Science of Climate Change, in *Public Understanding of Science*, 3(3): 259–275.
Beck, Silke (2004) Localizing Global Change in Germany, in Jasanoff, Sheila; Martello, Marybeth Long (eds.) *Earthly Politics. Local and Global in Environmental Governance* (Cambridge, MA: MIT Press), pp. 173–194.
Boehmer-Christiansen, Sonja (1994) Global Climate Protection Policy: The Limits of Scientific Advice. Part 1, in *Global Environmental Change*, 4(2): 140–159.
Boehemer-Christiansen, Sonja (1994) Global Climate Protection Policy: The Limits of Scientific Advice. Part 2, in *Global Environmental Change*, 4(3): 185–200.
Brand, Karl-Werner et al. (1997) *Ökologische Kommunikation in Deutschland* (Opladen).
Bray, Dennis and Hans von Storch (1999) Climate Science: An Empirical Sample of post-Normal Science, in *Bulletin of the American Meteorological Society*, 80(3): 439–455.
Bundesministerium für Bildung und Forschung (BMBF) (ed.) (2003) *Herausforderung Klimawandel. Bestandsaufnahme und Perspektiven der Klimaforschung* (Bonn).
Clark, William C. and Nancy Dickson (eds.) (1995) *The Press and Global Environmental Change: An International Comparison of Elite Newspaper Reporting on the Acid Rain Issue from 1972 to 1992* (Cambridge, MA: John F. Kennedy School of Government, Center for Science and International Affairs).
Dunlap, Riley E. (1998) Lay Perceptions of Global Risk. Public Views of Global Warming in Cross-National Context, in *International Sociology*, 13(4): 473–498.
Fogel, Cathleen (2004) The Local, the Global, and the Kyoto Protocol, in Sheila Jasanoff and Marybeth Long Martello (eds.) *Earthly Politics. Local and Global in Environmental Governance* (Cambridge, MA: MIT Press), pp. 103–126.
Heinrichs, H. (2003) Umweltrisiken und ihre sozio-kulturelle Verarbeitung: 'Bedeutungskonstruktion' am Beispiel der medienvermittelten Klima-Kommunikation, in A.Volkens et al., *Orte nachhaltiger Entwicklung: Transdisziplinäre Perspektiven* (Berlin: VÖW- Schriftenreihe), pp. 133–138.
Heinrichs, H. (2005) Kultur-Evolution. Partizipation und nachhaltige Entwicklung, in Gerd Michelsen and Jasmin Godemann (eds.) *Handbuch Nachhaltigkeitskommunikation. Grundlagen und Praxis* (München: Oekom-Verlag).
Kempton, Willet (1997) How the Public Views Climate Change, *Environment*, 39(9): 12–21.
Linneweber, Volker et al. (2001) *Soziale Repräsentationen von Entwicklungen in Natur- und Anthroposphäre auf Sylt vor dem Hintergrund globalen Wandels. Teilvorhaben der Fallstudie Sylt. Un veröffentlichter Schlussbericht des BMBF-Projekts 01 LK 9536/8* (Magdeburg).
Luhmann, Niklas (1986) *Ökologische Kommunikation* (Opladen).
Luhmann, Niklas (1996) *Die Realität der Massenmedien* (Opladen).
Mazur, Allan and Lee Jinling (1993) Sounding the Global Alarm: Environmental Issues in the US National News, *Social Studies of Science*, 23: 681–720.
McCright, Aaron M. and Dunlap, Riley E. (2000) Challenging Global Warming as a Social Problem: An Analysis of the Conservative Movement's Counter-Claims, *Social Problems*, 47(4): 499–522.
McCright, Aaron M. and Riley E Dunlap (2003) Defeating Kyoto: The Conservative Movement's Impact on U.S. Climate Change Policy, *Social Problems*, 50(3): 348–373.

Peters, Hans Peter (1994) Mass Media as an Information Channel and Public Arena, *Risk: Health, Safety & Environment*, 5(3): 241–250.

Peters, Hans Peter and Sippel, Marion (1998) Der Treibhauseffekt als journalistische Herausforderung, in Peter Borsch and Jürgen-Friedrich Hake (eds.) *Klimaschutz. Eine globale Herausforderung* (Landsberg am Lech: Aktuell), pp. 293–316.

Schirmer, M., and Schuchardt, B. (2005) *Klimawandel und präventives Risiko- und Küstenschutzmanagement an der deutschen Nordseeküste (KRIM): Synthesebericht* (Endbericht an BMBF).

Social Learning Group (2001) *Learning to Manage Global Environmental Risks, Vol. 1: A Comparative History of Social Responses to Climate Change, Ozone Depletion and Acid Rain; Vol. 2: A Functional Analysis of Social Responses to Climate Change, Ozone Depletion and Acid Rain* (Cambridge, MA: MIT Press).

Stehr, Nico and Hans von Storch (1995): The Social Construct of Climate and Climate Change, *Climate Research*, 5(2): 99–105.

Sterr, H. et al. (1999) Weltmeere und Küsten im Wandel des Klimas. *Petermanns Geographische Mitteilungen*, 143, 1999 / *Pilotheft 2000* (Justus Perthes Verlag Gotha), pp. 24–31.

Sterr, H. et al. (2000): Climate Change and Coastal Zones: An Overview of the State-of-the-Art on Regional and Local Vulnerability Assessment. Nota di Lavoro 38.2000 (Fondazione Eni Enrico Mattei. Mailand).

Trumbo, Craig (1996) Constructing Climate Change: Claims and Frames in US News Coverage of an Environmental Issue, in *Public Understanding of Science*, 5(3): 269–283.

Turner, R. K. et al. (1995): Assessing the Economic Cost of Sea Level Rise, *Environment and Planning* A 27: 1777–1796.

Weingart, P. (2001) *Die Stunde der Wahrheit? Zum Verhältnis der Wissenschaft zu Politik, Wirtschaft und Medien in der Wissensgesellschaft* (Opladen).

Weingart, P. et al. (2002) *Von der Hypothese zur Katastrophe. Der anthropogene Klimawandel im Diskurs zwischen Wissenschaft, Politik und Massenmedien* (Opladen).

Weinstein, Neil D. (1989) Optimistic Biases about Personal Risk, *Science*, 246: 1232–1233.

Wiedemann, Peter M. (1992) Klimaveränderungen: Risiko-Kommunikation und Risikowahrnehmung. In Peter Borsch and Peter M. Wiedemann (eds.) *Was wird aus unserem Klima?* (München: Bonn Aktuell), pp. 224–252.

Wilkins, Lee (1993) Between Facts and Values: Print Media Coverage of the Greenhouse Effect, 1987–1990, *Public Understanding of Science*, 2(1): 71–84.

Index

3G system, 185
9/11, 38, 48, 111, 115, 123–4

Accidents, 34
  accident theory, 55
Agenda, 21, 243
AIDS, 31, 149ff
  consequences of, 6
  crime associated with, 151
  denial, dissidents, 153, 154, 163
  effect on agriculture, 150
  effect on economy, 150
  and food security, 151
  gendered impact of, 152
  and human security, 151
  migration, 153,
  prevalence, 150
  role of media, 154–61
  and social breakdown, 151
  as threat to democracy, 151
  as transnational risk, 152–9
  Treatment Action Campaign, 153
Al-Qaeda, 54, 55, 115
  see also terrorism; terrorist networks
Annan, K., 26, 161
Armed conflict, 8
Asset welfare capitalism, 72
Austria, 82–3
Autism, *see* MMR vaccine

BBC *Watchdog Health Check*, 192, 193
Beck, U., 1, 5–6, 7, 32, 34, 94, 99
Beggar thy neighbour policy, 22
Bernoulli, D., 81
Bhopal, 38, 40
Biodiversity, 21
  loss of, 25
Blood transfusions, as health risk, France, 169
Bosnia, 102
Boundaries and borders, 27
Boycott, R., 195

Brown, M. D., 37
BSE, 38, 40, 69, 169, 170
  see also food safety; food-related risk
Building codes and standards, 51
Bush, B., 37, 41
Bush, G. W., 65

Central and Eastern European
  countries (CEEC)
  economic restructuring, 215
  pauperization and poverty in, 215, 216
  social deprivation, 212
  unemployment, 211, 216
  see also EU enlargement
Certainty, 3
Chance, 4–5, 29, 44, 94
Chemicals, 67–71
  downstream users, 69–70
  registration of new substances, 68
  see also pharmaceuticals; waste
    material risk assessment
Chernobyl, 4, 6, 38, 121, 122
  see also nuclear accidents; nuclear
    power; nuclear sites
Children's health, 168
  see also MMR vaccine; mobile masts
CIMIC, 104
Civil conflict, 101
Civil disobedience, 144
Civil liberties, infringement of, 98
Civil protection, 98
Civil society, 186
  global, 142–4
  responsibilities of, 161–3
Civil-military coordination, 105
  military mission creep, 104
Class society, 60
Climate change, 4, 11, 25, 60, 64, 229ff
  awareness, 236
  media reporting, 233–4, 235–45

253

Climate change – *continued*
  risk, 237–8
  see also coastal protection; extreme weather; greenhouse gas emissions; Hurricane Katrina; Kyoto Treaty; storm surges; tsunami
Climate policy, 63–5
Climate protection, international agreements, 242
Coastal protection, 229ff
  awareness, 236
  infrastructure, 242
  integrated coastal zone management, 231
  risk, 237–8
Coastal zones, 231
Coghill, R., 194
Cold War, 4
Common good, 61, 97
Communication, 12–14
  technology, 153
Communism, legacy of in Eastern Europe, 211
Compensation, 6
  see also liability
Conflict
  armed, 8
  civil, 101
  peaceful resolution, 9
  prevention, 100, 101
  violent, 101
Congo, 102
consequence management, 98, 121–2
Corporate capitalism, 48
Counter-terrorism, 98
Crisis response, 102
Critical events, 79
Cults, 52
Cultural diversity, 21
Cultural identity, 21
Culture, 48
  as dependent variable, 47
  tribal, 2

Damage
  limitation, 49
  prevention, 49
Danger, 4
  see also hazards

Decentralization, 56
Defence of the state, territory, 96
Deliberative learning, 43
Denmark, 74, 83
Dependency, 53
Design Basis Threat, 114–15
  see also nuclear accidents; nuclear power; nuclear sites, defence of
Development, 26–7
Development journalism, 160
Diethylstilbestrol, 171
Disaster
  deliberate, 52
  early warning systems, 35
  management, 243
  natural and manmade, 98
  prediction, forecasting, 37
Distribution of risk and harm, 42–3
Douste-Blazy, P., 176
Dworkin, R., 79

Eco-efficiency, 65–6
Ecological aggression, 24–6
Economic elites, 48
Efficiency, 56
  energy efficiency, 242
Ellwood, D., 42
Emergency warning systems, 51
Employment policy, 81
Endowment effect, 76
Ensuring state, 3, 8, 78
Environmental aftercare, 69
Environmental degradation, 65
Environmental diseases, 60
Environmental justice, 142
Environmental policy, 22–3, 59ff
Environmental risk, 9
European Employment Strategy, 73, 75
European identity, 210
European Security and Defence Policy, 103
European Security Strategy, 99, 100, 101, 102
European Social Model, 72
European Union, 61–2
  crisis prevention, 100–6
  eastern enlargement, 207ff
  emission guidelines, 193
  integrated coastal zone management, 231

European Union – *continued*
  Lisbon Strategy, 74
  policymaking, 102–3
  Programme on the Prevention of
    Violent Conflict, 101–2
Euroscepticism, 217
Extreme weather, 64, 229
  see also climate change; Hurricane
    Katrina; storm surges; tsunami

Financial crises, 60
First responders, 48
Food safety, 169
Food-related risk, 169, 170
Foot and mouth disease, 169
Frame, framing
  analysis, 41
  problem, 41
  theory, 76
France
  blood transfusion scandal, 169
  Hepatitis B vaccine crisis,
    175–8

Geertz, C., 2
Georgia, 102
Germany, 209
  as advocate of EU enlargement, 209, 210
  Basic Law, 61
  employment rates and ratios, 74
  Green Party, 8
  Hartz Commission, 84
  Rhine, environmental consequences of course modification, 22
  Zukunftsfähiges Deutschland, 66
Glaciers, melting, 239
Globalization, 8, 21
GM foods, 61, 66–7, 170
Governance, 7, 40
Greenhouse gas emissions, 9, 64, 238
  emission trading, 64
  mitigation, 242

Habermas, J., 12
Hacking, I., 29
Harm, 28, 29, 112
Hazards, 94
  identification, 39
  technological, 32

Health
  insurance, 11
  policy, 9
  risks, 167ff
  scandals, 169
  see also AIDS; BSE; France, blood transfusion scandal; hepatitis B vaccine, ionizing radiation; MMR vaccine; non-ionizing radiation
Health journalism, 162
Health-e, 162
Hepatitis B vaccine, 175–8
Herd immunity, 172
Homeland defence and security, 49
Houghton, J., 231
Human rights violation, 31, 101
Human Rights Watch, 143
Humanitarian aid, 104
Hurricane Katrina, 28, 36–7
Hyland, G., 188, 190, 194

Identity politics, 210
Immunization *see* hepatitis B vaccine; MMR vaccine
Income
  inequality, 50–1
  protection, Austria, 82–3
  see also poverty; wage compensation; wage insurance
Independent Expert Group in Mobile Phones, 184, 185, 186–7
Inequality, 42
Injustice, 5–6
Instability, 100
  regional, 101
Institutional state, 3
Insurance contract, 75
Integration argumentative, 219–20
Interdependency, 53
International Atomic Energy Agency, 112, 114
International Panel of Climate Change Reports, 230
Internet, 9–10, 53, 54, 56
Intervention, pre-emptive, 97
Ionizing radiation, 112, 120
  see also nuclear accidents
Iraq, 98

Israel, 98
  Six-Day War, 97
  see also pre-emptive strikes
Issue culture, 233

Job protection, 74, 78
Johannesburg Earth Summit, 62–3, 64

Kahnemann, D., 76
Kinkel, K., 213
Knowledge societies, 34
Kouchner, B., 176
Kyoto Treaty, Protocol, 47, 64, 65, 230

Labour market, 73–6
  transitional, 78–81
Languages, 21
Leake, J., 192
Lean production, 66
Liability, 6, 7, 23–4
Liberal democracy, 10
Life-course savings plan,
    Netherlands, 84
Loss, 4
  aversion, 78
  entailing events, 12
  loss/gain coefficient, 76
Low skilled, 73–4
Luhmann, N., 4

Maastricht Treaty, 61
Macedonia, 102
Mad cow disease, see BSE
Manhattan Project, 29
  see also nuclear holocaust; risk,
    incalculable
Marginalization, marginalized, 32, 37
Marx, K., 60
Mast Action UK, 185
  see also mobile masts; mobile
    phones
Measles, 172
  see also MMR
Media
  autonomy, 186
  climate change reporting, 233–4,
    235–45
  hepatitis B reporting, 176
  importance of, 186
  MMR reporting, 173
  mobile phone reporting, 185,
    189–91, 194
  nuclear hazard risk reporting, 124–6
  role in disseminating information
    about HIV/AIDS, 154–61
  scare stories, 125
Medicine, preventive, 4
  see also immunization
Microshield, 193
Microsoft, 51–2
Microwave health effects, 194
Millennium Development Goals, 24
Mitigation, 48
MMR vaccine, 172, 173
  association with autism, 173, 174
  association with inflammatory
    bowel disease, 173
  media coverage of, 172, 173, 174
  take-up, 172–3, 174
Mobile masts, 184, 185
  campaigns against, 188, 189, 190,
    195, 196–8
  and children, 188
Mobile phones, health risks of, 184ff
  EMF emissions, 185, 187, 188
  media reporting, 185–7
  see also mobile masts
Mobility insurance, 82
  in Denmark, 83
Modernity, 1
Modernization, reflexive, 32, 44
Moran, C., 190, 194, 195
Movement for environmental
    justice, 136
Movement for the Survival of the
    Ogoni People, 138
  see also Nigeria, Saro-Wiwa, K.
Multilateralism, 213, 214

Nagin, R., 36
National Commission on
    Terrorist Attacks upon the
    United States, 115
Natural and environmental disasters,
    28, 98
Neo-liberal shock therapy, 216
Netherlands, 51, 83, 84
New Orleans, 36–7

NGOs, 31
Nigeria, 131ff
  CHIKOKO, 140
  economy, 132
  ENGOs, 143
  ethno-linguistic
    groups, 131
  Federal Environmental Protection
    Agency, 137
  GDP, 132
  INC, 140
  Kaiama Declaration, 140
  Land Use decree, 139, 141
  MORETO, 140,
  MOSIEND, 140
  MOSOP, 140, 141
  NDDC, 137
  Ogoni Bill of Rights, 140
  Ogoni region, 136
  oil revenue, 132, 133, 135–6
  oil sector, risk management, 135
  OMPADEC, 137
  political structure, 131–2
  sabotage, 141
  pollution, 135, 139, 141
  state creation, 134, 136
  see also Saro-Wiwa, K.
Non-ionizing radiation, 188
North Sea coast, 232
  see also coastal protection
Nuclear accidents, 113
  see also Chernobyl
Nuclear energy
  information in public domain, 123–4
  opposition to, 126
Nuclear holocaust, 9, 29, 30
Nuclear power, 111ff
Nuclear sites
  defence of, 114
  proliferation, 111
  robustness, 120–1
  sabotage, 113
  security breaches, 118–20
  security regimes, 117
Nuclear terrorism, 111ff
  assessment, 114–17
  public perception of risk, 122–7
Nuclear weaponry, 29

Oil corporations, multinational, 137–8
Oil exploitation and exploration, 131ff
One World AIDS Radio network, 102
Opportunity, 80
Organized crime, 95, 101
Oxfam, 42

Panic, 48
Pauperization, 215
Peace-building, 9, 102
  role of military, 104
Peacekeeping, 104
Perrow, C., 7, 34
Pharmaceuticals, 171–2
  clinical trials, 171
Piot, P., 161
Plato, 3
Poland, 209
Policing
  as law enforcement, 105
  preventive, 98
  see also surveillance
Policy failure, 169
Pollutants, 39
  see also waster material risk assessment
Polluter pays principle, 6, 10
Population growth, 23
Postmodern state, 8
Post-trust society, 168, 170
Poverty, 26, 31
Power, 48
Precaution, 3, 10–11
  culture of, 3, 60–2
Precautionary principle, 47, 59ff, 96, 97
Pre-emptive strikes, 98
Prevention, preventive policy, 9–10, 47
Prisoner's dilemma, 76
Probability, 94
  probability theory, 81
Public opinion, 33
Public sphere, 186, 233
Public trust, 169

Radiation sickness, 112
Radioactive waste, 111

Index

Rational choice theory, 1, 76
Rationality, instrumental, 4
Rawls, J., 79
REACH, 67–70
Reid, J., 175
Reliability, 54–5
Relief work, 104
Renewable energy, 64, 242
Reynard, D., 188
Rio Declaration, 62
Rio Principles, 23
Risk
 amplification, 169
 analysis, 32, 33, 39
 assessment, 1, 29, 32, 39, 73
 communication, 40
 contamination, 169–70
 distribution, 5–6, 32
 estimation, 77–8
 global, 59
 governance, 31, 34–5, 40
 health-related, 167ff
 incalculable, 30, 33
 management, 31; proactive, 96–9
 as opportunity, 4–5
 perception, 77; gendered, 241
 policy, 43
 scientification, 12
 society, 1, 4, 32, 59
 spreading, 44
 tolerance, 167
 transnationalization, 7–8, 95
Risk-aversion, 73, 76, 77, 168
Risk-taking, 77, 78
Risky behaviour, 11
Risky choices, 73
Ritual magic, 2–3
Rule of law, 6, 105

Sabotage, 111, 113
Saro-Wiwa, K., 136, 138, 139, 140, 143
Schmidt-Bleek, F., 66
Science, trans-science, 33
Sea levels, 229, 231
Security
 international, 93, 94–6
 policy, 93ff
 preventive, 96, 97

 risk, 93
 terminology, 93
Sen, A., 79
Severance pay
 Austria, 82
 Netherlands, 83
Severe weather, 50
 see also climate change; extreme weather; storm surges
Shell, operations in Nigeria, 138, 139
Silicon Valley, 55
Size, and efficiency, 53–4
Social insurance, 7
Social risk management, 76, 78–81
Social wealth, 59
Society, stratified, 60
Sociology, organizational, 34
Solidarity, 72
Southern Africa, 149ff
Southern African Development Community, 143
Sovereignty, 8
 national, 99
Stability, 3, 212, 213
 regional, 102
 structural, 101
Stewart Group *see* Independent Expert Group in Mobile Phones
Stockholm Earth Summit, 22
Storm surges, 229, 240
Stress, 75
Surveillance, 98
Sustainable development, 61, 63, 142
Switzerland, 84
System theory, 4

Technology, 33–4
 high-risk, 39
Terrorism, 8, 39, 50, 95–6, 101
 assessment, 114
 containment, 13
 at nuclear facilities, 111ff
 see also al-Qaeda; counter-terrorism; nuclear sites; sabotage
Terrorist networks, 54, 55, 96
Thalidomide, 171
Third way, 72

Third World, dependency, 53
Transnationality, 95
Trust, 126, 245–8
Tsunami, Indonesian, 28, 35–6
  gendered impact of, 42
Tversky, A., 76

UK
  Anti-Terrorism, Crime and Security
    Act, 123
  control orders, 98
  New Deal programme, 74
  Prevention of Terrorism Act, 98
  Sellafield nuclear site, 117
  see also BSE; MMR vaccine; mobile
    phones; mobile masts
UNAIDS, 149
Uncertainty, 2–4
Unemployment, 74
  unemployment insurance, 79, 82
Uniformity, 21
United Nations, 21ff
  Charter, 24
  Environment Programme (UNEP),
    8, 21, 25–6, 63
  Global Media AIDS initiative, 154,
    161
US
  California, 51, 62
  Climate Stewardship Act, 62
  Florida, 50
  Food and Drug Administration, 171
  Louisiana, 51
  National Research Council, 39
  Nuclear Regulatory
    Commission, 114
  power grid, 53–4
  risk culture, 51
  see also 9/11; Hurricane Katrina
Utility, 76

Vulnerability, 41–2, 50, 56

Wage
  compensation, 84
  insurance, 83
Wakefield, A., 173
  see also MMR vaccine
Wall St game, 77
Waste material, risk assessment, 67
Water
  catchment basins, 23
  shortages, 60
Weinberg, A., 32–3
Weizsäcke, E. von, 66
Welfare state, 7, 8, 12, 72
Work
  capacity, 75
  precarious, 74
  see also income inequality; job
    protection; unemployment
Working career, compressed, 75
Work-life insurance, 79–80, 82
World Trade Organization, 24